# Writing Scientific English

A textbook of English as a
Foreign Language for students of
Physical and Engineering Sciences

## John Swales

## Nelson

*Uniform with this title*:
Posner: Practice in English
Bruton: Exercises on English Prepositions and Adverbs

*Also*:
The Bruton English Course:
    Students' Books 1–3
    Teachers' Books 1–3
    Language Laboratory Drills 1–3
    Recorded Material.

710219838 -8

# Contents

# Preface

I have attempted in this book to outline and give practice in a series of structured contexts through which foreign students of physical sciences and engineering can develop their ability to express their scientific and technical knowledge in English. In writing this course I have had two particular groups of students in mind. Firstly, it should be useful for students who are required to use English in the course of their higher studies in science and engineering, but whose knowledge of the language is limited to what they have learnt in a general school course—a course likely to have concentrated on spoken English and more 'literary' kinds of comprehension and composition, and to have dealt little, if at all, with the language of science and technology. Secondly, it should be of use and interest to students who are specialising in science in their last years at school. Obviously it is at school that preparation in the language skills necessary for coping with English-medium, or partly English-medium, scientific and technical instruction must be begun. I hope, therefore, that this course will do something to resolve the continuing crisis in the field of teaching non-native speakers of English how to communicate scientific and technical information in the most widely used language of science, because it is precisely in this area that the lack of *relevant* language preparation holds up so many students in their university and technical college work.

The scope of this book has been restricted in a number of ways. First, it is primarily designed to improve *written* English. I have excluded comprehension work because I have found that the way in which a comprehension passage is best handled depends principally on how much the students know of the scientific subject-matter. For this reason I believe that the selection and presentation of comprehension passages should be carried out within an actual teaching situation.

Secondly, I have restricted the 'register' to that of intermediate scientific text-books, and ignored features more typical of articles in journals. I have also tried to maintain a clear distinction between 'popular' journalistic writing on science topics and simple Scientific English. Similarly, I have limited the types of exercise to those requiring the student to produce no more than fairly short and fairly straightforward descriptions, explanations and interpretations. The work in this book, therefore, stops short of instructing the student how to write technical reports, reviews, discussions or theses.

Thirdly, I have tried to isolate the main grammatical difficulties likely to be encountered in descriptive work and give practice in them, both at an elementary and intermediate level. This has led to one or two unexpected omissions. The reader will not find, for example, any detailed discussion of Conditional Clauses. This is because the preliminary analysis suggested that the hypothetical and counter-factual types of condition are not, in fact, either typical or useful at the level of straightforward description. The 'simple' Conditional Clause is needed, but the factors affecting the choice of Present Simple or 'will' are highly complex and not fully understood at the present time. I have also decided that the customary distinction between 'defining' and 'non-defining' Relative Clauses is, for teaching purposes, arbitrary and largely irrelevant in so far as any study of scientific writing immediately throws up many examples of relative clauses that appear to be neutral in this respect.

Fourthly, the content of the examples and exercises has been limited to the fields of physics, chemistry and engineering because I feel that rather different syntactic and organisational difficulties confront foreign students of biology, agriculture, zoology, etc.

Finally, I have on many occasions proposed fairly precise ways of organising different kinds of written work. I do not want to imply that the procedures outlined are the only—or indeed the best—ways of approaching certain writing tasks. These models of sentence and paragraph organisation have been drawn up to help students, especially weaker students, to write more correctly, more coherently, and more within the style best suited to scientific and technical subjects.

I would like to thank all those colleagues at the University of Libya who tried out provisional drafts of the material with their classes. However, I owe a greater debt to Hugh Mildmay, who helped in so many ways, and to my wife for her continuing encouragement and critical interest.

John Swales,
The Institute of Education,
The University of Leeds.

**Author's note**

The exercises have been marked ○, △, or □.

○ exercises are simple and should give little difficulty
if the explanations and examples have been studied
carefully.

△ exercises usually require students to produce a
certain amount of their own work. However, quite
a lot of help is given in terms of example sentences
and in the organization of the written material.

□ exercises are rather more advanced and nearly always
require students to produce passages of continuous
scientific or technical English.

# Unit 1  Introduction to scientific statements

## Be and have in scientific statements

In scientific English the main verbs of sentences are usually in the Present Simple tense. It is not difficult to see why. Scientific textbooks contain information about the present state of scientific knowledge. They describe experiments showing how this knowledge can be obtained. They also show how this knowledge is used in the service of man. As a result, you will probably use the Present Simple in most of your scientific writing.

As the Present Tense is so common, it is important to make sure that the subjects and verbs agree.

○ **Exercise 1**  Underline the subjects of these sentences and cross out the verbs which do *not* agree. (Underline all parts of the subjects, not just the nouns.) Here is an example:

*This gas has/have a greater density than air.*
*This gas has/~~have~~ a greater density than air.*

1 Water boils/boil at 100° centigrade.
2 Action and reaction is/are opposite and equal.
3 A thermometer measures/measure temperature.
4 Oxygen and hydrogen is/are gases.
5 Mathematics is/are an important subject for an engineer.
6 The light bulbs in this room produces/produce 100 watts each.
7 The liquid in those bottles is/are dangerous.
8 The results of the experiment proves/prove the law.
9 Everybody recognizes/recognize the importance of practical work.
10 On average, women lives/live longer than men.
11 Some substances, most of which are metal, is/are good conductors of electricity.
12 Most kinds of wood floats/float on water.
13 At least one kind of wood sinks/sink in water.
14 The average monthly rainfall figures for this area shows/show a small decline in annual total over the last thirty years.
15 The apparent loss of weight of a substance which is immersed in a liquid equals/equal the weight of the displaced liquid.

The last two sentences (14 and 15) are examples of an important fact about scientific writing: the main verb is often simple but the rest of the sentence complicated. In 14 and 15 the verbs consist of only one word. How many words do the subjects contain?

Compare sentences 14 and 15 with these two spoken sentences:

*'I don't want to go and see him today.'*
*'What are you going to do tomorrow?'*

In spoken English the noun parts of a sentence are often simple and the verb parts complicated. The opposite is true of written scientific English. In fact, about a third of all scientific statements have *is* or *are* as the main verb. This causes difficulty for students who speak languages in which it is not always necessary to use a verb like *be*.

The difficulty arises because English is one of those languages in which all written sentences must contain at least one main verb.

○ **Exercise 2** Rewrite these 15 sentences putting in the main verb *is* or *are*. (This is the first writing exercise in the book. Therefore you should try to get it right. This also means:

> starting each sentence with a capital letter
> finishing each sentence with a full stop
> making sure you copy the words accurately.)

1 These test-tubes.
2 Cast-iron not as strong as steel.
3 Oxygen necessary for all growth.
4 Oxygen and hydrogen gases.
5 Oxygen, like hydrogen, a gas.
6 This solution a mixture of chlorine and sodium.
7 Angles measuring 90° right-angles.
8 The natural water in many parts of the world hard.
9 Gold and silver not radio-active elements.
10 One of the machines out of order.
11 Two of the three pieces of metal copper.
12 A beaker or a small glass necessary for this experiment.
13 The spiral motion of air above a low-pressure area always opposite in direction to the movement of the hands of a clock.
14 The breaking strain of the rope 200 kilos.
15 200 kilos the breaking strain of the rope (Be careful!)

The other very common verb in scientific statements is the main verb *have*. Again this can cause a problem because of the grammatical differences between English and many other languages.

○ **Exercise 3(a)** Put either *is* or *has* into the spaces.

1 Water ........ a boiling point of 100° C.
2 The boiling point of water ........ 100° C.
3 Stainless steel ........ a metal alloy.
4 Stainless steel ........ rust-proof.
5 This car ........ a maximum speed of 140 kilometers an hour.
6 The maximum speed of this car ........ 140 kilometers an hour.
7 The angle of reflection ........ 9°.
8 The simplest hydrocarbon ........ methane which ........ one carbon atom and four hydrogen atoms.
9 In chemistry each element ........ its own symbol, which ........ usually a capital letter followed by a small letter.
10 If a plane figure ........ three straight sides, it ........ a triangle.

△ **Exercise 3(b)** Write five scientific statements using *has*:

1 giving the freezing point of a liquid
2 giving the melting point of a metal
3 giving the density of a substance
4 giving a property of a square
5 giving a property of a triangle

○ **Exercise 4** In the following sentences the main verbs have been left out. Rewrite the sentences putting in either *is*, *are*, *has*, or *have*.

1 A triangle a figure which has three straight sides.
2 The Dead Sea a high salt content.
3 There several types of pump.
4 Most kinds of stainless steel a small percentage of chrome.
5 Stainless steel the property of resisting corrosion.
6 Modern bridges often several kilometers long.
7 A modern bridge sometimes a length of several kilometers.
8 Isosceles triangles two equal angles.
9 The total population of the world about 3,500 million.
10 A hexagon a plane figure with six sides.

Read this description of a car.

## The Moto 1100

*The Moto 1100 is a small family car. It has a small engine which is in the front. The engine has a capacity of 1,100 cubic centimeters. It is a front wheel drive car. The gear lever is on the floor. There are seats for four or five people. It has four forward gears and a reverse. It has a maximum speed of about 130 km an hour. One advantage of this car is that it has a very low fuel consumption.*

Notice how it is possible to write a simple technical description using only *be* and *have*. Notice also that the sentences are short.

△ **Exercise 5**   Write a simple factual description of any vehicle you know about. Use mainly *be* and *have*. Keep most of your sentences short. (A *vehicle* is any car, bicycle, lorry, etc.)

## Statements requiring the Present Simple

(a)  The Present Simple is used for regular actions and regular processes:

*He studies physics six hours a week.*
*The crude oil then passes down the pipe-line.*

(b)  It is used for general statements:

*Water freezes at $0°$ C.*
*Area equals length times height.*

(c)  It is used for factual statements and observations:

*This type of vinegar contains about 3% acid.*
*The liquid in the test-tube weighs 55 grams.*

(d)  It can be used in descriptions of experiments:

*The filter paper then collects the impurities.*
*The temperature rises until it reaches $100°$, but after that it remains constant.*

In other words, always use the Present Simple unless there are good reasons for using another tense.

### Form of the Present Simple

$$\left.\begin{array}{l} it \\ he \\ she \end{array}\right\} \; produce + \mathbf{S} \qquad\qquad \left.\begin{array}{l} they \\ we \\ you \\ I \end{array}\right\} \; produce$$

A large **S** has been used for two reasons. First it is a reminder, so that you remember to add it on after a subject in the third person singular.

A large S has also been used because it is not always a simple matter of adding an *s* to the base form of the verb. Sometimes spelling changes are necessary:

Verbs ending in *ss*, *sh*, *ch*, *x*, and *o* add *es* to the base.

| | |
|---|---|
| *they pass* | *he passes* |
| *they push* | *he pushes* |
| *they watch* | *he watches* |
| *they mix* | *he mixes* |
| *they go* | *he goes* |

Verbs ending in *y* after a consonant change *y* to *i* and add *es*.

| | |
|---|---|
| *they hurry* | *he hurries* |
| *they magnify* | *it magnifies* |

Verbs ending in *y* after a vowel follow the main rule (add *s*.)

| | |
|---|---|
| *they obey* | *it obeys* |
| *they say* | *he says* |

○ **Exercise 6** Rewrite these sentences putting the verbs in brackets into the correct form.

1 He (study) biology.
2 The current (pass) along the wire.
3 This ring (weigh) 125 grams.
4 Sound (travel) at a speed of 333 meters a second.
5 Rain (wash) salt from the soil.
6 This factory (employ) thirty people.
7 This bird (catch) insects as it (fly).
8 Water (solidify) or (turn) into ice at 0° C.
9 Glue (fix) or (stick) two surfaces together.
10 The down-stroke of the piston (compress) the mixture until it (explode).

○ **Exercise 7** Write out the third person singular of the following verbs.

| | | | |
|---|---|---|---|
| 1 study | 6 analyze | 11 convert | 16 lay |
| 2 reach | 7 do | 12 possess | 17 employ |
| 3 try | 8 design | 13 mix | 18 qualify |
| 4 think | 9 apply | 14 assemble | 19 draw |
| 5 stretch | 10 weigh | 15 exchange | 20 supply |

5

△ **Exercise 8** The ten sentences below have been mixed up. Study them and write out the ten correct sentences, choosing one item from each column each time. Do not use any item more than once. (−) means that no preposition is required.

| | | | |
|---|---|---|---|
| Fresh water | freezes | at | 78° C. |
| Alcohol | equals | about | an average of 70 years. |
| Copper | liquefies | at | 10 per cent alcohol. |
| Oxygen | weighs | − | 0·9104 meters. |
| Wine | consists | at | −183° C. |
| A yard | contains | at | one kilogram per litre. |
| Sound | live | − | 2 oxygen atoms and 1 carbon atom. |
| Mercury | travels | of | 1083° C. |
| Men | melts | for | 760 mph at sea level. |
| A carbon dioxide molecule | boils | at | −39° C. |

○ **Exercise 9** Rewrite putting the verbs in brackets into the correct forms.

Around the earth there (be) a large area of gas which (form) the atmosphere. This layer of gas—or more accurately, gases- (provide) some of the chemical materials which man (need). The other raw materials (come) from the earth and the sea.

About four-fifths of the atmosphere (consist) of nitrogen. The remainder (be) mostly oxygen. The other five gases (be) very rare and, in fact, (make up) less than 1 per cent of the total atmosphere. Although these gases (be) rare, at least two of them (have) common uses in the field of electrical lighting. Electric light bulbs usually (contain) argon. Neon (be) also useful because it (give) out light when an electrical current (pass) through it.

○ **Exercise 10(a)** Rewrite the passage below putting in the six verbs which follow it. Use each verb-form once only. Make sure that the verbs agree with the subjects.

### Colour

When sunlight strikes an object the colour of the object depends upon the wavelengths which the object ........ . If, for example, grains of sugar ........ equally all the wavelengths of the spectrum the grains ........ white. If a surface reflects only the wavelength which produces red and ........

the other waves of the spectrum, the surface ....... red. Black is the absence of colour because black objects .... ... all the light of the spectrum.

*absorb*    *absorbs*        *appear*    *appears*        *reflect*    *reflects*

△ **Exercise 10(b)** Write sentences in answer to these questions:

1 What colour surface absorbs the greatest amount of heat?
2 What does a mirror do?
3 How fast do light rays travel?
4 Some rays are invisible—give at least two examples.

△ **Exercise 11** The figures show a force pump in operation. The statements which follow describe the two stages of the cycle. However, the sentences are in the wrong order. Study the diagrams, decide the correct order of the sentences and write them out in two continuous passages. (If you like, also use linking-words like *first*, *then*, *next*, etc.)

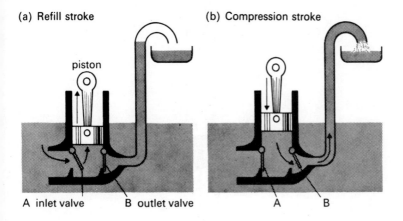

(a) Refill stroke    (b) Compression stroke

piston

A inlet valve  B outlet valve   A  B

*Refill stroke*

1 A mechanical force pulls the piston upwards.
2 The reduced pressure causes the inlet valve (*A*) to open.
3 Water enters the cylinder and fills the space beneath the piston.
4 The outlet valve (*B*) remains closed because of the pressure of water in the outlet pipe.
5 The pressure in the cylinder beneath the piston begins to fall.

7

## Compression stroke

1 Water continues to flow up the tube until the piston reaches its lowest point.
2 A mechanical force pushes the piston downwards.
3 The cycle starts again.
4 This causes the inlet valve to close.
5 The high pressure water is forced up the outlet tube.
6 The cylinder now traps the water because it cannot escape back through *A*.
7 When the pressure is sufficiently high the outlet valve opens.
8 The pressure beneath the piston begins to rise.

□ **Exercise 12** Write a short description of how the syringe works. (You may be able to get some help from the previous exercise.)

Syringe

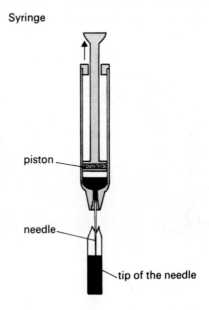

piston

needle

tip of the needle

Negative statements and questions are much less common in written scientific English than in other varieties of English. With regard to the Present Simple, the important thing to remember is that both negatives and questions are formed with the help of the verb *do*, which is 'empty' of meaning.

| statement | negative | question |
|---|---|---|
| *Wood floats* | *Iron does not float* | *Does wood float?* |
| *Sticks burn* | *Stones do not burn* | *Do stones burn?* |

○ **Exercise 13** If you think that certain of the following statements are not true, write them out as negative statements. If you think they are true, leave them.

1 Iron floats on water.
2 The opposite angles of a parallelogram equal 180°.
3 Water boils at 100° centigrade.
4 A piece of iron corrodes when it is buried under the ground.
5 Parallel straight lines meet.
6 A gold coin corrodes when it is buried under the ground.
7 Steel melts at 600° centigrade.
8 The weight of a metal varies with the temperature.
9 In a right-angled triangle the length of the hypotenuse equals the sum of the lengths of the other two sides.
10 A scientist uses the Present Continuous tense when he writes about observable facts.
11 A barometer enables scientists to measure pressure.
12 We make drills and cutting tools from low carbon steel.
13 The volume of a gas varies with its pressure.
14 At the top of its stroke the piston comes into contact with the cylinder cover.
15 A black surface reflects sunlight well.
16 An electric shock of a hundred volts usually causes death.
17 The magnetic needle of a compass points towards the South Pole.
18 Aluminium dissolves in water.
19 A mixture of concentrated nitric and hydrochloric acid (1 to 4 by volume) dissolves gold.
20 Sentence 17 contains a spelling mistake.

△ **Exercise 14** Look at this example:

*A man leaves A at 10.00 A.M. for B which is 4 km from A. He walks at 6 kph.*
*When does he arrive at B?*

Now write suitable questions for the following:

1 A train leaves *A* at 10.00 A.M. for *C*, 20 km away. It arrives at 10.25 A.M.
   (a) How fast ........?

B

2  A train leaves $A$ for $D$, which is 35 km away. It travels at an average speed of 70 kph.
   (a) When ........?
   (b) How long ........?
3  A train leaves $A$ for $E$. It travels at 80 kph and arrives at $E$ one hour and twenty minutes later.
   (a) How far ........?
4  $x$ weighs 2·5 kilos a liter and $y$ weighs half as much as $x$.
   A can weighing half a kilo contains 6 liters of $y$.
   (a) How much ........?
5  Make up a problem about distance/speed/time.
6  Make up a problem about weight/cost.
7  Make up a problem about length/width/area.

If your problems are in correct English perhaps the teacher will try and work out the answers.

# Unit 2  Dimensions and properties

## Dimensions

The dimensions of an object are its length, height, volume etc.

*A solid has three dimensions.*
*A surface has two dimensions.*
*A line has one dimension.*
*A point has no dimensions.*

There are several different ways of describing dimensions in English. The simplest way—and the most common way outside the field of science and technology—uses *be* as the main verb. We will call it structure 1. Here are some examples:

|  | *be* |  |  | *adjective* |
|---|---|---|---|---|
| *x* | *is* | *3* | *centimeters* | *long.* |
| *The mountain* | *is* | *2,150* | *meters* | *high.* |
| *The river* | *is* | *50* | *meters* | *wide.* |
| *The well* | *is* | *45* | *meters* | *deep.* |
| *The pipes* | *are* | *4·5* | *centimeters* | *thick.* |
| *The river* | *is* | *50* | *meters* | *broad.* |

You will meet two possible spellings of words like *meter* and *centimeter*. The *-er* spelling (which is American) is used here rather than the *-re* spelling (which is British). The *-er* spelling is easier and more logical.

*Wide* and *broad* usually mean the same thing and can be used one instead of the other. This can be seen in the two statements about the river.

*Tall* and *high* also mean the same thing, but they cannot usually be used one instead of the other.

*Tall* is used of physical objects which are much longer in height than in width.

11

*High* is used of rounder or squarer objects. Therefore we say *a tall tower* but *a high dam*.

Also notice that only *high* can be used when describing things which are not physical objects: *a high speed, high pressure* etc.

○ **Exercise 1** Complete these sentences using a suitable adjective. The first one has already been done.

1 The mountain is 2,150 meters ........
2 The carpet is 3 meters ........
3 The carpet is 1½ centimeters ........
4 The chimney is 15 meters ........
5 The telephone wires are 6 meters ........
6 The telephone poles are 6·5 meters ........
7 That tree is 20 meters ......
8 The box is 0·50 meters ........
9 Women are usually about 1·50 meters ........
10 The door is 7 feet ........ , 2½ feet ........ , and 2½ inches ........

○ **Exercise 2** Here are some 'notes.' Write them out in full using structure 1. Here is an example:

*this table—2 meters*
*This table is 2 meters long.*

1 this ruler—30 centimeters
2 this ruler—3 centimeters
3 this ruler—0·3 centimeters
4 standard writing paper—0·1 millimeters
5 this river—15 meters
6 the river—2 meters—in the middle
7 the lake—7 kilometers—and 14 kilometers
8 size 16 nails—2 centimeters—and 0·2 centimeters
9 telephone poles—usually 7 meters—and 20 centimeters
10 recent types of transistor—only 0·4 centimeters
11 the pages of this book ....
12 the opposite wall ....
13 the nearest window ....
14 Mount Everest ....
15 the local railway track ....

△ **Exercise 3** Write two short passages, one giving the dimensions of the box, the other the dimensions of the building. Continue to use structure 1. (Diagrams on the next page)

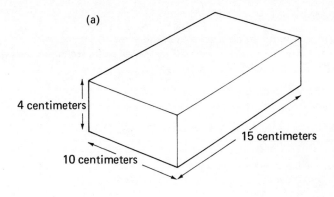

(a)

4 centimeters

15 centimeters

10 centimeters

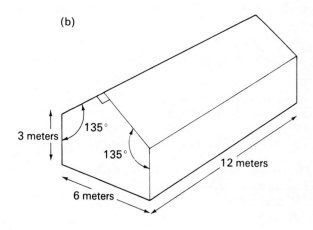

(b)

3 meters

135°

135°

12 meters

6 meters

The other main way of giving dimensions uses the verb *have*. We will call it structure 2. Here are some examples:

|  | *have* |  | *noun* |  |
|---|---|---|---|---|
| *x* | has | a | length | of 3 centimeters. |
| *The mountain* | has | a | height | of 2,150 meters. |
| *The river* | has | a | width | of 50 meters. |
| *The well* | has | a | depth | of 45 meters. |
| *The pipes* | have | a | thickness | of 20 centimeters. |

Notice these points:

(a) The indefinite article is used with the noun of dimension.
(b) The noun from *tall* (*tallness*) is never used in this structure.
Therefore, the two correct sentences are:

*This man is 1·75 meters tall.*
*This man has a height of 1·75 meters.*

(c) With structure 2 it is now possible to make statements about how heavy something is:

*This stone has a weight of 85 grams.*
*This type of car has a weight of 950 kilograms.*

○ **Exercise 4** Complete these sentences. Use structure 2. Here is an example:

*This pen .... 11 centimeters.*
*This pen has a length of 11 centimeters.*

 1 The road outside .... 6 meters.
 2 The tallest building in the street .... 12 meters.
 3 This pen .... 0·8 centimeters.
 4 The contents of the test-tube .... 250 grams.
 5 The top-soil in this area .... 15 centimeters.
 6 The sample .... 9 kilos.
 7 The walls of the glass container .... 15 millimeters.
 8 This room .... 7 meters and .... 4 meters.
 9 3 liters of water ....
10 1 liter of petrol ....

These 'dimension' nouns are difficult to spell. Look at the following list, cover *A* and choose one of the five spellings in *B* for each of the nouns. Then check them against the correct spellings in *A*. If you make a mistake write the word out correctly at least three times.

| A | B | | | | |
|---|---|---|---|---|---|
| *length* | longth | length | lengeth | lenght | lenegth |
| *height* | height | hight | heigth | hieght | heihgt |
| *width* | wideth | weidth | width | widht | wiedth |
| *depth* | depeth | deeph | debth | depth | depght |
| *weight* | wieght | weight | weigth | weght | waight |

Study these abbreviations (short forms):

| | | | | |
|---|---|---|---|---|
| *fig.* | = | *figure* | *kg* | = *kilogram(s)* |
| *km* | = | *kilometer(s)* | *gm* | = *gram(s)* |
| *m* | = | *meter(s)* | *mg* | = *milligram(s)* |
| *cm* | = | *centimeter(s)* | *hr* | = *hour(s)* |
| *mm* | = | *millimeter(s)* | *min.* | = *minute(s)* |
| *approx.* | = | *approximately* | *sec.* | = *second(s)* |

Notice that the same abbreviation is used for singular and plural. But remember that a plural form such as *3 cm* must be pronounced as *3 centimeters*.

Finally, there are a few other nouns of dimension that have not been mentioned so far. The most common ones are:

*radius, diameter, area, volume, circumference*

○ **Exercise 5** Complete these sentences using the information given:

*Circle A has a radius of 3 cm.*

1 Circle *A* .... 6 cm.
2 Circle *A* .... 18·84 cm.
3 Circle *A* .... 28·26 cm².

*Box B: length = 3m; height = 1·5 m; width = 2 m.*

4 Box *B* .... 9 m³.
5 Box *B* .... 27 m².

*Sphere C has a diameter of 10 cm.*

6 Sphere *C* .... 5 cm.
7 Sphere *C* .... 31·4 cm.
8 Sphere *C* .... 528 cm³.

*Cylinder D has a cross-sectional area of 28·26 cm² and a height of 12 cm.*

9 Cylinder *D* ....
10 Cylinder *D* ....

△ **Exercise 6** Describe the dimensions of the car and the steel pipe. Write a short passage on each. In (b) include a statement about the amount of steel used in the pipe. (In order to avoid repeating the same structure all the time it is a good idea to use both structures 1 and 2.) (Diagrams on the next page)

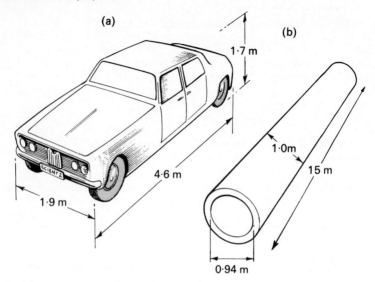

## Properties

We have already seen that *have* can be used to make statements of length, height and weight etc. *Have* is also commonly used in referring to many other properties.

○ **Exercise 7** Here are ten mixed-up sentences. Study them and write out the ten correct sentences.

| | | | |
|---|---|---|---|
| Aluminium | has | a speed of | 15 years. |
| Water | has | an average life of | 0·000026 per °C. |
| Glass | has | a range of | 100° C. |
| Sound | have | a resistance of | more than 5,000 km. |
| Cows | has | a coefficient of expansion of | 2·8. |
| Some modern planes | has | a specific gravity of | −114° C. |
| Alcohol | has | a capacity of | 760 mph at sea level. |
| The storage tank | have | a boiling point of | 2,500 ohms per 100 cm. |
| This wire | has | a freezing point of | 5,000 liters. |

16

These *have* sentences have one thing in common; they all describe properties. Consider this pair of sentences:

*Water boils at 100° C.*
*Water has a boiling point of 100° C.*

The first sentence states that some action (boiling) takes place at a certain temperature. However, it may be more scientific not to think of water actually doing something (in this case, boiling), but of having certain properties such that certain things occur at a certain temperature. This is why the second sentence may be preferred even though it is longer.

△ **Exercise 8** Complete ten of the following, using sentences of your own:

1  .... (*have*) a boiling point of .... .
2  .... a freezing point of .... .
3  .... a density of .... .
4  .... a velocity of .... .
5  .... a mass of .... .
6  .... a voltage of .... .
7  .... a breaking strain of .... .
8  .... an average life of .... .
9  .... an average temperature of .... .
10  .... a resistance of .... .
11  .... a diameter of .... .
12  .... a force of .... .
13  .... a cost of .... .
14  .... a thermal conductivity of .... .
15  .... the property of .... .

There are several possible ways of putting structure 2 statements in the negative. Look at the following:

Common in spoken British English

*Pure water hasn't got a smell.*
*Pure water hasn't got any smell.*

Common in spoken American English

*Pure water doesn't have a smell.*
*Pure water doesn't have any smell.*

17

Common in written scientific English

*Pure water has no smell.*
*A point has no dimensions.*

☐ **Exercise 9** Look at the table below. You will see that it is not complete, but you probably know some of the missing information. Write as much as you can about each of the six substances. (Notice that you cannot use *have* when describing the form.) Join some of the statements together. Here is an example:

*Aluminium is a metal which has a melting point of 660° C.*
*It is silver in colour and it has a specific gravity of 2·8.*
*It has no smell.*

| Substance | Form | Colour | Smell | Melting Point | Boiling Point | Density |
|-----------|------|--------|-------|---------------|---------------|---------|
| chlorine | gas | | | | | 0·0032 |
| oxygen | | none | none | −218 | −183 | 0·0014 |
| ethyl-alcohol | | | character-istic | | 78·5 | |
| water | | | | | | |
| iron | | | | 1535 | 2800 | |
| sulphur | solid | yellow | | 113 | 445 | 2·07 |

## 'Fronted' statements

Look at these examples of 'fronted' statements (structure 3):

|  | noun |  | be |
|------|------|--------------|--------------|
| The | length | of x | is 3 cm. |
| The | width | of this river | is 50 m. |
| The | height | of the hill | is 750 m. |
| The | depth | of this well | is 45 m. |
| The | thickness | of this tree | is 1½ m. |

Notice the difference between structure 2 and structure 3 (the subjects have been underlined):

structure 2    The well *has a depth of 45 meters.*
structure 3    The depth of the well *is 45 meters.*

In structure 3 more information is put into the subject. This way of building up the subject is common in scientific and technical writing. It can be called 'fronting' because more words are put in front of the main verb. Here are some more examples:

*The specific gravity of benzene is 0·78.*
*The distance between the two contacts is 2·5 mm.*
*The coefficient of expansion of brass is 0·000026 per ° C.*
*The diameter of the cheaper kinds of household electric wire is approximately 1·3 mm.*

△ **Exercise 10** Rewrite these sentences using structure 3. Here is an example:

*This pen is 11 cm long.*
*The length of this pen is 11 cm.*

1 The bottle weighs 160 gm.
2 Water freezes at 0° C.
3 The water-towers have a height of 35 m.
4 Sound has a speed of 333 m per sec.
5 The water-tower has a capacity of 50,000 gallons.
6 The sea 100 m from the shore has an average depth of 15 m.
7 The cheaper kinds of household electric wire are 1·3 mm across.
8 The two cylinders of oxygen are 38 kg and 41 kg in weight.
9 The temperature in the furnace averages 900° C.
10 Under these circumstances, gravity has no effect.

*Summary of structures for stating dimensions and properties*

structure 1:   *x is 3 cm long.*
structure 2:   *x has a length of 3 cm.*
structure 3:   *The length of x is 3 cm.*

It is also possible to describe dimensions using a variation of structure 1:

structure 1(a):   *x is 3 cm in length.*

□ **Exercise 11** Write a passage describing the lay-out of a football field. (Be careful not to confuse the three structures.) (Diagram on the next page)

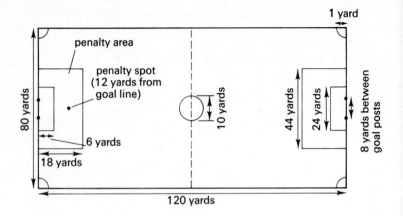

## Qualified statements of dimensions

If a dimension is not given exactly, the fact that it is not exact should be made clear. In non-scientific English we usually use the word *about*:

*x is about 3 centimeters long.*

In more technical writing *approximately* may be used instead:

*x is approximately 3 cm long.*
*x is approximately 3 cm in length.*

Here are some other typical qualifying words and phrases:

(a)  $x = 3\ cm$     *x is 3 cm long.*
(b)  $x = 3·00\ cm$     *x is exactly 3 cm long.*
(c)  $x = \pm 3\ cm$     *x is about 3 cm long.*
                      *x is approximately 3 cm long.*

(d)  $x1 = 3\ cm$
     $x2 = 3\ cm$
     $x3 = 2·8\ cm$ } *x is 3 cm long on average*
     $x4 = 3·2\ cm$

(e)  $x = 3·3\ cm$     *x is over 3 cm long.*
(f)  $x = 2·7\ cm$     *x is under 3 cm long.*
(g)  $x = 3·1\ cm$     *x is slightly over 3 cm long.*
                      *x is just over 3 cm long*
                      *x is a little over 3 cm long.*
(h)  $x = 2·9\ cm$     *x is just under 3 cm long*
                      *x is a little under 3 cm long.*
                      *x is slightly under 3 cm long.*

What structure was used in the examples you have just read?

Of course the above examples are only approximate. It is not possible to say exactly when these qualifying words or phrases are to be used.

**Exercise 12** Write ten sentences as indicated below. Here is an example:

*The length of AB = 9·03 cm (just over)    AB is just over 9 cm long.*

1  $x = 3·07$ cm long
   (a) (exactly)
   (b) (approximately)
2  The value of $\pi = 3·14159$
   (a) (approximately)
   (b) (slightly under)
   (c) (to two decimal places)
3  The width of the pipe = 0·216 meters
   (a) (under)
   (b) (just over)
   (c) (exactly)
4  The speed of the plane = 523 kph
   (a) (very approximately)
   (b) (a little over)

The qualifying phrases *under, over, just under, a little over,* etc. can be used in the same way with structures 2 and 3:

structure 2   *x has a length of over 3 cm.*
                 *x has a length of slightly under 3 cm.*
                 *x has a length of a little over 3 cm.*

structure 3   *The length of x is over 3 cm.*
                 *The length of x is slightly under 3 cm.*
                 *The length of x is a little over 3 cm.*

However, there are two possible forms with other qualifications:

structure 2   *x has an approximate length of 3 cm.*
                 *x has a length of approximately 3 cm.*

structure 3   *The approximate length of x is 3 cm.*
                 *The length of x is approximately 3 cm.*
                 *The average length of x is 3 cm.*
                 ....
                 *The exact length of x is 3 cm.*

△ **Exercise 13** Rewrite these sentences qualifying them.

1  The mountain is 2045 m high.
2  The height of the mountain is 2045 m.
3  The foundations of the building are 3·9 m deep.
4  The samples have weights of 18·6, 21·1, and 19·5 kilograms.
5  The nerve is 0·009 mm in breadth.
6  Diamond has an index of refraction of 2·47.
7  The moon has a radius of 1736 km.
8  Light has a speed through water of 224 million meters per second.
9  Under stated conditions atmospheric pressure equals 14·72 psi.
10  The escape velocity of the moon is 2·38 km per sec. and for the earth 11·2 km per sec.

□ **Exercise 14** Describe the dimensions of this syringe. Include a statement about the approximate amount of liquid the syringe can contain.

Scale 1 : 3

# Unit 3  Comparisons and modals

## Simple statements of comparison

Study these example sentences:

$x > y$
1  *x is longer than y.*
2  *A meter is longer than a yard.*
3  *In summer the desert is hotter than the coast.*
4  *Soft woods are cheaper than hard woods.*
5  *X-rays are shorter in wavelength than light rays.*
6  *Hydrochloric acid is more dangerous than citric acid.*
7  *Benzene is more complicated chemically than methane.*
8  *A diesel engine is more efficient than a steam engine.*
9  *River-water usually contains more impurities than well-water.*
10  *Sea-water contains more manganese than fresh water.*

Notice these points:

(a)  Short adjectives usually take *-er* and *-est.*
(b)  Long adjectives usually take *more* and *most*\*.
(c)  *More* and *less* are used with singular nouns:

*Arts students usually do less work than science students.*
*If this happens, add more water to the mixture.*

(d)  *Fewer* is used with plural nouns:

*Fewer oil-wells will be drilled this year.*

Notice that as $x > y = y < x$, so:
*x is longer than y  =  y is shorter than x.*

○ **Exercise 1** Change the ten example sentences into the reverse structures. The first one has already been done.

---

\* Adjectives of medium length—more precisely of two syllables—take *-er*, or *more*,
or either. Which adjective takes which comparative form is very complicated and
perhaps it is best to use the form you find in scientific text-books. As a general
rule, it is 'safer' to use *more* or *most*. To give an idea of the difficulty we can
write *abler* but not *stabler, commoner* but not *moderner.*

The main verb *have* is also commonly used in making statements of comparison. These are frequently in the form of structure 2, as in the following examples:

*Alcohol has a lower boiling point than water.*
*A hexagon has more sides than a quadrilateral.*
*A hexagon has a greater number of sides than a quadrilateral.*
*Figure* (b) *has a smaller area than Fig.* (a)

○ **Exercise 2(a)** Make statements of comparison about the two figures. The first one has already been done.

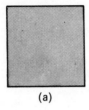 

(a)  (b)

1 A comparison of area.
2 A comparison of the number of sides.
3 A comparison of the length of sides.
4 A comparison of the size of angles.
5 A comparison of the number of angles.

○ **Exercise 2(b)** Make statements of comparison about the two buildings.

 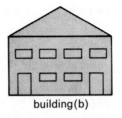

building (a)  building (b)

1 A comparison of the number of doors.
2 A comparison of the number of windows.
3 A comparison of the height of building.
4 A comparison of the number of storeys.
5 A comparison of the height of roof.

24

○ **Exercise 3** The eight sentences below have been mixed up. Study them and write out the eight correct sentences.

| Water | has | a higher carbon content | than | Mexico. |
| Gases | has | a higher octane rating | than | a cube of similar base area. |
| A pyramid | have | a higher boiling point | than | ordinary petrol. |
| Alaska | has | a greater coefficient of expansion | than | mild steel. |
| Iron | has | a smaller volume | than | ethylalcohol. |
| Oil wells | has | a greater specific gravity | than | solids. |
| Hard steel | has | a greater depth | than | aluminium. |
| Aircraft fuel | have | a lower average temperature | than | water wells. |

△ **Exercise 4** Here is some information about two cars. Write a short passage comparing the two. Remember to begin with a general statement and then discuss the detailed differences. Notice that you may use either *be* or *have*:

*Have* { Car B has a higher price than Car A.
{ Car A has a lower price than Car B.

*Be* { Car B is more expensive than Car A.
{ Car A is cheaper than Car B.

| | Car A | Car B |
|---|---|---|
| Price | £1,000 | £1,300 |
| Engine size | 1,000 cc | 1,500 cc |
| Fuel consumption | 7 lit./100 km | 9 lit./100 km |
| Length | 4·2 m | 4·6 m |
| Height | 1·7 m | 1·6 m |
| Maximum speed | 130 kph | 145 kph |

Notice these spelling changes:

(a) Adjectives ending in consonant+y take *-ier* and *-iest* in the comparative and superlative. Hence: *Heavy, heavier, heaviest.*
(but *grey, greyer, greyest*: why?)
(b) Adjectives ending in single vowel letter + single consonant letter double the last letter. Hence: *hot, hotter, hottest.*
(but *weak, weaker, weakest*: why?)

c

○ **Exercise 5** Cross out the wrong comparative forms:

| | | | | | |
|---|---|---|---|---|---|
| 1 | biger | bigger | 6 | slower | slowwer |
| 2 | shortter | shorter | 7 | widder | wider |
| 3 | deepper | deeper | 8 | taller | taler |
| 4 | fatter | fater | 9 | nearrer | nearer |
| 5 | closer | closeer | 10 | windyer | windier |

If there are more than two things to be compared, we can write:

*x is longer than a, b, or c*  or  *x is the longest.*

Notice that with the verb *be* there are two possibilities:

(a)  i  *The kilometer is the longest metric unit of measurement.*
     ii *The longest metric unit of measurement is the kilometer.*

(b)  i  *Mount Everest is the highest mountain in the world.*
     ii *The highest mountain in the world is Mount Everest.*

(Which are the two fronted sentences?)

○ **Exercise 6** Answer these questions in complete sentences.

1 Which is the longest metric unit of measurement?
2 Which is the shortest metric unit of measurement?
3 Which is the heaviest metal?
4 Which is the lightest gas?
5 Which is the simplest atom?
6 Which is the most complicated atom?
7 Which is the biggest city in the world?
8 Which is the least expensive metal?
9 Which are the most important oil-producing countries?
10 Which subject do you find most difficult and why?

In a similar way, both:

*Car D is the most expensive.*
*The most expensive car is D.*

are possible. However, scientists may prefer the second sentence because it shows fronting. In addition, scientists may prefer to use a noun rather than an adjective and write:

*The car with the highest price is D.*

This kind of sentence can be called 'noun-fronted.' Now consider 'noun-fronted' sentences used to describe dimensions:

*The car with the longest length ....*
*The car with the biggest length ....*
*The car with the most length ....*
None of these three is really correct. Can you see why?

The correct form is: *The car with the greatest length ....*

△ **Exercise 7** Answer these questions about the four cars using information supplied in the table. Use noun-fronted structures as much as possible, e.g., *The car with the highest price is D.* The first one has been done.

|  | **Car A** | **Car B** | **Car C** | **Car D** |
|---|---|---|---|---|
| Price | £1,000 | £1,300 | £650 | £2,000 |
| Engine size | 1,000 cc | 1,500 cc | 750 cc | 1,900 cc |
| Fuel consumption | 7 lit./100 km | 9/100 km | 6/100 km | 11/100 km |
| Maximum speed | 130 kph | 145 kph | 115 kph | 170 kph |
| Length | 4·2 m | 4·6 m | 3·8 m | 5·0 m |
| Height | 1·7 m | 1·6 m | 1·6 m | 1·7 m |
| Number of passengers | 4 | 5 | 4 | 6 |

1 Which car is the most expensive?
2 Which car is the cheapest?
3 Which car is the longest?
4 Which cars are the lowest?
5 Which car has the biggest engine?
6 Which car can go the fastest?
7 Which car uses the least petrol?
8 Which car can seat the most people?
9 Which car has the least space for passengers?
10 Which car uses the most petrol?

There is another way of making comparisons which is commoner in scientific and technical English than in other kinds of writing. Look at these examples:

*y is shorter than x = y is not as long as x.*
*Mathematics is more interesting than English = English is not as interesting as Mathematics.*

The use of this structure with compared nouns needs care:

*Water has a higher boiling point than alcohol = Alcohol does not have as high a boiling point as water.*
*Kuwait has more oil than Syria = Syria does not have as much oil as Kuwait.*
*A polygon has more sides than a triangle = A triangle does not have as many sides as a polygon.*

○ **Exercise 8** Transform the following into *not...as...as* sentences. The first one has already been done.

1 $x$ is longer than $y$.
2 A meter is longer than a yard.
3 Soft woods are cheaper than hard woods.
4 A diesel engine is more efficient than a steam engine.
5 Tungsten has a higher melting point than platinum.
6 River-water usually contains more impurities than well-water.
7 Sea-water contains more manganese than fresh water.
8 Copper is a better electrical conductor than aluminium.
9 Fewer students will fail this year.
10 The Dead Sea has a higher salt content than any other sea.

△ **Exercise 9(a)** Study the table and then write ten sentences comparing the time spent by the two groups doing various things. Here are three examples:

*Group 2 does not spend as much time sleeping as Group 1.*
*Group 2 spends fewer hours sleeping than Group 1.*
*Group 1 spends more time on hobbies and entertainment than Group 2.*

**How two groups of students spend their week**

|  | Group 1 | Group 2 |
| --- | --- | --- |
| Sleep | 64 | 60 |
| Classes | 23 | 20 |
| Study | 12 | 24 |
| Travelling | 8 | 10 |
| Eating | 16 | 12 |
| Hobbies | 12 | 8 |
| Entertainment | 15 | 10 |
| Sport | 4 | 7 |
| Other activities | 14 | 17 |
| Total | 168 hours | 168 hours |

□ **Exercise 9(b)** Write a paragraph comparing how you spend your week with that of the students in Group 2. (Be honest!)

## Qualified comparative statements

Comparative statements can be qualified in order to make them more informative—exactly how much bigger?, approximately how much longer?, and so on.

Study these example sentences. Notice that the comparatives have been qualified.

*A meter is slightly longer than a yard.*
*A meter is a little longer than a yard.*
*A yard is considerably longer than a foot.*
*A yard is much longer than a foot.*
*A yard is three times longer than a foot.*
*A meter is more than three times longer than a foot.*
*Line AB is 3 cm longer than CD.*
*Benzene is somewhat more complicated chemically than methane.*
*A diesel engine is at least 50 per cent more efficient than a steam engine.*
*Sea-water contains considerably more manganese than fresh water.*

Again notice that:

*A meter is slightly longer than a yard = A yard is slightly shorter than a meter.*

○ **Exercise 10** Change the ten example sentences above into the reverse structures. The first one has already been done.

The meaning of these qualifying words is shown approximately in this table:

| | | |
|---|---|---|
| ▨ | A is the same size as B | ▨ |
| ▨ | A is minimally larger than B | ▨ |
| ▨ | . . . slightly larger/a little larger . . . | ▨ |
| ▨ | . . . somewhat larger . . . | ▨ |
| ▨ | . . . rather larger . . . | ▨ |
| ▨ | . . . much larger/considerably larger . . . | ▨ |
| ▨ | . . . very considerably larger . . . | ▨ |

Certain qualifications are also possible with the *as...as* structure. It is mainly used with *almost* and *nearly* and with *times – twice as..., three times as..., half as...,* etc.

*A meter is a little longer than a yard* = *A yard is nearly as long as a meter.*
*A meter is a little longer than a yard* = *A yard is almost as long as a meter.*
*A yard is three times longer than a foot* = *A yard is three times as long as a foot.*
*A meter is more than 3 times longer than a foot* = *A meter is more than 3 times as long as a foot.*

Some qualifications of the *as...as* structure become rather complex and need not be used for the moment. For instance, the approximate equivalent of:

*x is very considerably longer than y*
is
*y is nowhere near as long as x.*

△ **Exercise 11** Complete ten of the following using sentences of your own. Think before choosing *be* or *have*.

1 .... not as heavy as .... .
2 .... nearly as tall as .... .
3 .... twice as far as .... .
4 .... almost as expensive as .... .
5 .... not as hot as .... .
6 .... not as pure as .... .
7 .... half as long as .... .
8 .... many times the speed of .... .
9 .... over twice as high as .... .
10 .... double the thickness of .... .
11 .... almost as useful as .... .
12 .... approximately three times the density of .... .
13 .... twice as much rainfall as .... .
14 .... approximately twice the range of .... .
15 .... nowhere near as complicated as .... .

△ **Exercise 12** The bottles contain varying amounts of water. Write the ten comparative statements about the five bottles. Here are two examples:

*d* and *e*   *Bottle d contains twice as much water as bottle e.*
*e* and *d*   *Bottle e contains half as much water as bottle d.*

| | |
|---|---|
| 1 *c* and *b* | 6 *a* and *b* |
| 2 *b* and *c* | 7 *b* and *a* |
| 3 *c* and *e* | 8 *a* and *d* |
| 4 *e* and *c* | 9 *d* and *a* |
| 5 *d* and *c* | 10 *c* and *a* |

(a) 1¼ litres  (b) 4 litres  (c) 12 litres  (d) 6 litres  (e) 3 litres

If there is no difference, these are some of the forms that are commonly used:

*x and y have the same length.* (structure number ....?)
*x and y are the same in length.* (structure number ....?)
*The lengths of x and y are the same.* (structure number ....?)
*The length of x is the same as the length of y.*
*The length of x is the same as that of y.*

△ **Exercise 13** Write ten statements based on the information below. Do not use the same structure for both members of a pair. Here is an example:

A————B  *AB and CD have the same length.*
C————D  *The lengths of AB and CD are the same.*

1  (a) (b)
   (a) (b)

2  *x* = 50 mph   (a)
   *y* = 80 kph   (b)

31

3  $x$ = 550 ft-lb wt/sec.          (a)
    $y$ = 7·5 $10^6$ cm-gm wt/sec.      (b)

4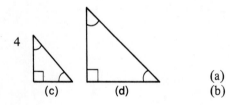

        (c)        **(d)**          (a)
                                          (b)

5  $M1$ = 15 ft/sec./sec.          (a)
    $M2$ = 15 ft/sec./sec.          (b)

Notice that we say:

*x equals y*    or    *x is equal to y*

We do not say:

*x equals to y*    or    *x is equal y.*

Notice that we say:

*He compared x and y*    or

*He made a comparison between x and y*

We do not say:

*He compared between x and y.*

Finally notice these two qualifications of superlative expressions:

*Nitrogen is easily the most common element in the atmosphere.*
*Nitrogen is by far the most common element in the atmosphere.*

Both *easily* and *by far* are special expressions. *Easily* has nothing to do with *easy* (the opposite of *difficult*) and *by far* has nothing to do with distance.
*London is easily/by far the biggest town in Britain.* (In other words, all the others are very much smaller.)

□ **Exercise 14** Study the graphs and the table, and then write a passage comparing the climates of cities *A* and *B*.

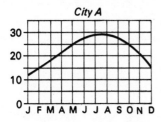

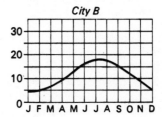

Average Monthly Temperature (° centigrade)

**Average Monthly Rainfall** (in millimeters)

|  | City A | City B |
|---|---|---|
| January | 82 | 47 |
| February | 46 | 39 |
| March | 37 | 47 |
| April | 5 | 38 |
| May | 3 | 41 |
| June | 1 | 52 |
| July | 0 | 63 |
| August | 0 | 61 |
| September | 9 | 45 |
| October | 39 | 68 |
| November | 67 | 60 |
| December | 92 | 61 |
| Total | 381 mm | 622 mm |

□ **Exercise 15** In fact city *B* is London. Look up the facts of temperature and rainfall about a city you know and write a paragraph comparing its climate with that of London.

## A note on modals in scientific English

After the Present Simple, the most common verb forms in scientific English are those which contain modals. The most frequent modals are:

group 1: *can, may, might, could*
group 2: *will*
group 3: *should, must, have to*

33

Modals are used with the base form of the verb to give extra meaning to the sentence. In spoken English it is very difficult to say exactly what these extra meanings are. In scientific English it is easier.

The modals in group 1 are frequently used to make statements of possibility and probability. Consider:

*The glass bottle breaks when dropped.* (Every time this type of bottle is dropped from this height onto this surface it breaks: approximately 98–100% probability of its breaking.)
*The bottle can break when dropped.* (A good chance that it will break: approximately 40–70% probability.)
*The bottle may break when dropped.* (Some chance that it will break: approximately 20–40% probability.)
*The bottle could/might break when dropped.* (A small chance: approximately 5–20% probability.)
*The bottle cannot break when dropped.* (Almost no chance: 0–2% probability.)

Notice that *cannot* is written as one word.

Notice also that *might* amd *could* are not past tenses of *may* and *can*.

○ **Exercise 16** Complete these sentences qualifying the main verbs with modals. The first one has been done.

1 Death occurs if the body temperature rises to 44° C.
   (a) Death *can occur* if the body temperature rises to 43° C.
   (b) Death .... if the body temperature rises to 42·5° C.
   (c) Death .... if the body temperature rises to 42° C.
2 If the pressure rises above 500 gm/cm$^3$, structural failure occurs.
   (a) If the pressure rises above 400 gm/cm$^3$, structural failure .... .
   (b) If the pressure rises above 300 gm/cm$^3$, structural failure .... .
   (c) If the pressure does not rise above 200 gm/cm$^3$, structural failure .... .
3 The new college will open in five years' time.
   (a) The new college .... in four years' time
   (b) The new college .... in three years' time.
   (c) The new college .... in thirty months' time.
   (d) The new college .... in five months' time.

*Will* (group 2). Compare:

(a) *The sea-water corrodes the iron.*
(b) *The sea-water will corrode the iron.*

(a) is a statement about the action of sea-water in a certain area on a certain piece of iron. On the other hand, (b) is a prediction about the action of certain sea-water on a certain piece of iron. (You predict when you claim that something will happen, as in, *It will rain tomorrow.*) Therefore, (a) and (b) have different meanings.

Now compare:

(c) *Sea-water corrodes iron.*
(d) *Sea-water will corrode iron.*

(c) is a general statement based on knowledge of the laws of science. (d) is a prediction based on knowledge of the laws of science. However, as making a statement about something that always happens and predicting it come to the same thing, (c) and (d) have the same meaning. Hence, *will* and the Present Simple have the same meaning in well-known, general scientific statements. This is why it is possible to write both:

*If pure water is heated to 100° C at sea level, it boils.*
*If pure water is heated to 100° C at sea level, it will boil.*

But in doubtful situations the difference between stating (Present Simple) and predicting (*will*) is maintained:

(e) *Sea-water does not corrode this new alloy* (may be wrong.)
(f) *Sea-water will not corrode this new alloy* (to be proved.)

○ **Exercise 17** Study these sentences. In some of them the modal can be replaced by the Present Simple, without any change in meaning. Rewrite these sentences in the Present Simple and leave the others.

1  The sea-water will corrode the iron.
2  Sea-water will corrode iron.
3  If alcohol is heated to 78° C it will boil.
4  Sea-water will not affect this new alloy.
5  Death will occur if body temperature exceeds 43° C.
6  A compass needle will always point towards magnetic north.
7  If the company drills in this area they will find the rock very hard.
8  Unless a certain critical temperature is reached the structure of the steel will not change.
9  If the satellite manages to return to the earth with the photographs, they will provide a great deal of information.
10  If small children drink only water they will suffer from calcium deficiency.

The most useful modal in group 3 is *should*. *Should* is often used in written warnings and instructions:

*Students should be careful when using acids.*
*Concrete should contain at least 12% cement.*
*Boiler pressure should not exceed 300 psi.*

○ **Exercise 18(a)**  List one thing a student *should do* and one thing a student *should not do*:

1  in a chemistry laboratory
2  when working with electricity
3  when writing scientific English

△ **Exercise 18(b)**  Write instructions for one of the following, using *you should...* as the main verb form. Also use *first, then, next*, etc., to make your instructions clearer:

1  How to give an injection
2  How to write up an experiment
3  How to address a letter
4  How to find a book in a library
5  How to draw a 60° angle with a pair of compasses.

# Unit 4 Impersonal scientific statements-the passive

## Form of the passive

All finite passives are formed by some part of the verb *be* plus the past participle.

Present passive $\left.\begin{array}{l} is \\ are \end{array}\right\}$      + past participle

| | | |
|---|---|---|
| *The gas* | *is* | *heated.* |
| *The bridge* | *is* | *made of concrete.* |
| *The gas* | *is carefully* | *heated.* |
| *Bridges* | *are usually* | *made of reinforced concrete.* |

Notice the position of the adverbs.

The modal passive $\left\{\begin{array}{l} will \\ can \\ may \\ should \\ etc. \end{array}\right\}$ + *be* + past participle

| | | | |
|---|---|---|---|
| *The survey* | *will* | *be* | *completed next year.* |
| *The survey* | *may* | *be* | *completed next year.* |
| *Acids* | *should* | *be* | *handled with great care.* |
| *Heat* | *can* | *be* | *generated in several ways.* |

## Negative statements and questions in the passive.

| | | | |
|---|---|---|---|
| *Litmus paper* | *is not* | *used* | *in this experiment.* |
| *His research* | *will not be* | *completed* | *this year.* |
| *Agriculture* | *cannot be* | *developed* | *without sufficient water.* |

Notice that *cannot* is written as one word.

| | | | | |
|---|---|---|---|---|
| | *Is* | *tea* | | *grown* *in Iran?* |
| | *Can* | *atoms* | *be* | *seen* *under a microscope?* |
| *Where* | *are* | *dams* | *usually* | *built?* |

## Spelling rules

(a) The past participle of regular verbs is formed by adding -*ed* to the base form, or by adding only -*d* to the base form if the verb ends in *e*.

| | | | |
|---|---|---|---|
| *heat* | *heated* | *cause* | *caused* |
| *boil* | *boiled* | *close* | *closed* |

(b) Verbs ending in consonant + *y* change to *i*.

| | |
|---|---|
| *apply* | *applied* |
| *carry* | *carried* |
| *occupy* | *occupied* |

(c) Verbs ending in vowel + *y* follow the normal rule.

| | | | |
|---|---|---|---|
| *delay* | *delayed* | *employ* | *employed* |

(d) Verbs ending in a single vowel + single consonant follow a rather complicated rule:

i Verbs of three or more syllables are regular.

| | | | |
|---|---|---|---|
| *develop* | *developed* | *deposit* | *deposited* |

ii Verbs of only one syllable double the final consonant.

| | |
|---|---|
| *plot* | *plotted* |
| *slip* | *slipped* |
| *stir* | *stirred* |

iii Verbs of two syllables double the final consonant if the second syllable carries the stress.

unstressed second syllable

| | | | |
|---|---|---|---|
| '*answer* | *answered* | '*cover* | *covered* |

stressed second syllable

| | | | |
|---|---|---|---|
| *pre'fer* | *preferred* | *ad'mit* | *admitted* |

This last rule has no exceptions in American spelling. In British spelling, however, verbs ending in single consonant + *l* double the *l* although the stress is on the first syllable.

| | |
|---|---|
| *level* | *levelled* |
| *chisel* | *chiselled* |

However, many of the commonest and most useful verbs in scientific English have irregular past participles.

○ **Exercise 1** Complete this list of irregular past participles.*

| | | | | |
|---|---|---|---|---|
| 1 break | broken | 4 bend | bent |
| choose | ........ | build | ........ |
| drive | ........ | burn | ........ |
| freeze | ........ | deal with | ........ |
| get | gotten/got | feel | felt |
| give | ........ | keep | ........ |
| shake | ........ | leave | ........ |
| take | ........ | lose | ........ |
| write | ........ | mean | ........ |
| 2 draw | drawn | 5 feed | fed |
| fly | ........ | hold | ........ |
| grow | ........ | lead | ........ |
| know | ........ | meet | ........ |
| show | ........ | read | ........† |
| 3 cut | cut | 6 tear | torn |
| put | ........ | wear | ........ |
| run | ........ | bring | brought |
| set up | ........ | think | ........ |
| | | 7 do | ........ |
| | | say | ........ |
| | | find | ........ |
| | | see | ........ |
| | | make | ........ |
| | | understand | ........ |

## Use of the passive

It is probable that in any physics, chemistry or engineering text-book at least one-third of all the finite verbs will be in the passive. Most of these passive verbs will be either in the Present Simple or be used with modals like *will, can, may* or *should.* Therefore, it is clear that scientists and engineers use the passive much more frequently than most other kinds of writer. Why is this?

First, study these pairs of sentences:

(a) *The litmus paper is placed in the liquid* (passive).
*He places the litmus paper in the liquid* (active).

---

* A complete list can be found on page 82.
† But pronounced *red.*

(b) *The gas is carefully heated* (........).
   *The experimenter heats the gas carefully* (active).
(c) *The results will be analyzed* (........).
   *Scientists will analyze the results* (........).
(d) *This is shown in Fig. 3* (........).
   *I show this in Fig. 3* (........).
(e) *Filaments are made of tungsten wire* (........).
   *People(?) make filaments of tungsten wire* (........).
(f) *A barometer is used for measuring atmospheric pressure* (........).
   *People(?) use a barometer for measuring atmospheric pressure* (........).

Although the two sentences in each pair have similar meanings, the passive sentences are clearer. The first reason for this is that the passive sentences do not mention people. For a scientist many references to people are unnecessary and confusing.

In (a), (b) and (c) for example, the mention of *He, The experimenter* or *Scientists* does not give the reader any useful new information. It is true that the passive sentences do not actually state who the people are, but it can almost certainly be guessed who they are from earlier sentences in the passages.

In (d) the use of *I* gives too much importance to the person and not enough to what he is doing. Compare these two sentences:

*I now weigh it.* (Brilliant, original stroke of scientific genius!)
*It is now weighed.* (The next and obvious step in the process.)

In (e) and (f) it is not clear what the subjects of the active sentences should be—this is why *people* is followed by (*?*). In fact, do people or machines make filaments? Or both? Do people actually use barometers? If so, who are these people? The writer? The writer and his readers? All educated people? Everybody? (Unfortunately, our small daughter used ours as a drum!) All these confusing questions can be avoided by using the passive.

A second reason for using the passive is that the subject is a very important part of the sentence. (Remember 'fronting' and how scientists often put a great deal of information into the subject.) Compare these two short passages. The subjects are in italics.

(a) *The experimenter* fixes a long metal bar in a retort stand by one end. *He* heats the other end in a flame until it becomes red. *He* then moves the bar and the stand away from the heat. *He* finds that the temperature of the hot end of the bar falls rapidly.

(b) *A long metal bar* is fixed in a retort stand by one end. *The other end* is heated in a flame until it becomes red. *The bar and the stand* are then moved away from the heat. *The temperature of the hot end of the bar* is found to fall rapidly.

In (a) the subjects of the sentences contain very little information and are repeated in an unhelpful and uninteresting way.
In (b), on the other hand, the subjects contain a lot of information. This information catches the reader's attention because it comes first in the sentence. Also notice how the steps in the experiment are *contrasted*:

*A long metal bar*
*The other end*
*The bar and the stand*
*The temperature of the hot end of the bar*

A third reason for using the passive is that passive sentences may be a little shorter. Look back at the six pairs of sentences. How many of the passive sentences are shorter than the active ones?

In many situations, therefore, the passive can be used to give the necessary information in the best possible way; impersonally, concisely, objectively, and giving importance to the most important facts. As a final reminder of the form, notice the grammatical relationship between:

*He* | places | | the litmus paper | *in the solution.*

| The litmus paper | | is placed | *in the solution.*

**Exercise 2** Read this passage, underline the passives and then answer the questions.

As oil is usually found deep in the ground its presence cannot be determined by a study of the surface. Consequently, a geological survey of the underground rock structure must be carried out. If it is thought that the rocks in a certain area contain oil, a 'drilling rig' is assembled. The most obvious part of a drilling rig is a tall tower which is called 'a derrick.' The derrick is used to lift sections of pipe, which are lowered into the hole made by the drill. As the hole is being drilled, a steel pipe is pushed down to prevent the sides from falling in, and to stop water filling the hole. If oil is struck a cover is firmly fixed to the top of the pipe and the oil is allowed to escape through a series of valves.

1 How many Present Simple passives are there in the passage?
2 Write down the two modal passives.
3 Write down the one Present Continuous passive.
4 Write down the two *plural* Present Simple passives.

D

5 Write down the two passives which include adverbs.
6 Why cannot the presence of oil be determined by looking at the ground?
7 What must be done before work is started?
8 What happens when a likely place for oil is found?
9 What is the metal tower called?
10 What is the derrick used for?
11 How is the entry of water prevented?
12 Why is a cover fixed to the top of the pipe?

△ **Exercise 3(a)** Answer at least seven of these questions.

1 Where is coffee grown?
2 Where are peanuts grown?
3 What is vegetable oil used for?
4 What are barometers used for?
5 In which parts of the world is mineral oil found?
6 What is the study of rocks called?
7 What is a magnetic compass used for?
8 What is the study of plane and solid figures called?
9 What are pipes usually used for?
10 Where is snow found?

△ **Exercise 3(b)** Complete at least seven of the following questions. Do not use the same questions as in the two previous exercises.

1 What .... called?
2 Where .... grown?
3 What .... used for?
4 What .... made of?
5 How .... prevented?
6 How .... used?
7 How .... calculated?
8 When .... needed?
9 Where .... found?
10 When .... finished?

△ **Exercise 3(c)** Make the following passage more technical by making it impersonal. To do this you will need to use the ........

First I take a small electric bell and hang it with its dry battery cell from a rubber band. I choose rubber because it doesn't transmit sound easily. Next I hang the bell inside a bell-jar. Then I place the bell-jar on a pumping table. As I pump the air out, the sound of the bell becomes fainter and fainter. You will see the hammer moving and striking the bell but you won't hear any sound.

Many passives are followed by prepositional phrases. Here are some common types.

*To*

*The temperature is then reduced to −240° C.*
*The results are given to three decimal places.*

In these sentences *to* means *as far as*. Notice the difference between:

*The liquid is cooled to 5° C.*
*The liquid is cooled 5° C.*

*From*

This is often used after verbs that have the general meaning of *get* or *obtain*. Here are three examples:

*Fresh water can be distilled from sea-water.*
*Domestic electricity is taken from power lines.*
*Paper is made from wood.*

*As*

This preposition frequently follows a passive verb containing the modal *can*. This is shown in these examples:

*The speed of sound can be taken as 333 m/sec.*
*This can be expressed as a curve.*
*A small mirror can be used as a reflector.*
*This is usually known as the Law of Inertia.*

△ **Exercise 4** Complete ten of these passive sentences. Don't forget the verb *be*:

1  .... distilled from sea-water.
2  .... extracted from ore.
3  .... removed from the heat source.
4  .... taken as 3·142
5  .... heated to 100° Centigrade.
6  .... known as relative density.
7  .... deduced from a knowledge of the current and voltage.
8  .... obtained from crude oil.
9  .... isolated from external sources.
10  .... taken as 39·3 inches.
11  .... reduced to $\frac{1}{3}$ of its original volume.
12  .... expressed as $(x + y)(x - y)$.

13 .... derived from the sum of the squares on the other two sides.
14 .... derived from a Greek word.
15 .... derived from an Arabic word.

△ **Exercise 5** Here are some 'notes.' Write them out in full. You must supply articles (where necessary), prepositions and the correct form of the verb. (You will need to use other prepositions in addition to *from*, *to* and *as*.) Here is an example:

*iron—can—write—symbolically—Fe.*
*Iron can be written symbolically as Fe.*

1 Mathematics—use—many different purposes
2 decimal system—use—everywhere—scientific writing
3 breaking strain—know—1%
4 for this experiment—force—gravity—can—regard—nothing
5 value—should—give—three decimal places
6 results—will—show—form—graph
7 modern planes—equip—jet engines
8 polystyrene—now—use—insulation—heat and cold
9 in this furnace—air—blow—molten copper—remove—impurities
10 *An Experiment to show that there is no resistance in a vacuum*
   feather and lead weight—fix—top—closed tube: tube—then—
   evacuate: tube—then—place—upright position: feather and weight—
   release: it—find—that—feather and weight—reach—bottom—tube—
   same time.

Hare's Apparatus

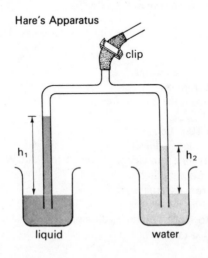

**Exercise 6** Explain how Hare's apparatus can be used to find the specific gravity of a liquid. (Diagram on last page).

## By and the agent

You may have noticed that up till now there has been no mention of *by*. There are two good reasons for this. Firstly, it is very important to realise that *by* + agent is not necessarily a part of a passive sentence. Secondly, a large majority of passive sentences in scientific English have no agent. In fact, *who*-agents (i.e., *who* it was done by) are rare and mainly used with the past tenses. *What*-agents are mainly used with the Present Tense and Modals. This can be seen in the following exercises.

**Exercise 7** Complete these sentences using the Present passive. Here is an example:

*Cars (drive) petrol engines.*
*Cars are driven by petrol engines.*

   1  Many car engines (cool) water.
   2  The Volkswagen car engine (cool) air.
   3  Expansion (cause) heat.
   4  Heat (generate) friction.
   5  Modern aeroplanes (power) jet engines.
   6  Sound waves (make) vibrations.
   7  Iron and steel (attract) magnets.
   8  Hydrogen (produce) the action of acid on zinc
   9  Whole numbers (form) a combination of the digits 0 to 9.
  10  Light and heat (reflect) the surface of the glass.

**Exercise 8** Complete ten of the following. Don't forget *be*.

   1  .... cooled by air.
   2  .... powered by steam.
   3  .... required by industry.
   4  .... eliminated by a filter-paper.
   5  .... found by calculation.
   6  .... separated by chromatography.
   7  .... controlled by a thermostat.
   8  .... measured by a voltmeter.
   9  .... separated by a prism.
  10  .... corrected by spectacles.
  11  .... supplied by a bunsen burner.
  12  .... affected by oxidation.

13 .... found by Ohm's law.
14 .... found by Charles' law.
15 .... determined by the number of electrons in its outermost shell.

Now consider the agents in these three statements:

*Many car engines are cooled by water.*
*The area of a circle can be found by calculation.*
*Impurities are eliminated by a filter.*

These agents refer to general principles; they give a general idea of how something is done. If it is necessary to explain a process in more detail the agents can be expanded:

*Many car engines are cooled by circulating water through the engine block.*
*The area of a circle can be found by using the formula $\pi r^2$.*
*Impurities are eliminated by passing the liquid through a filter.*

These *what*-agents can be made even more detailed:

*Many car engines are cooled by circulating water through the engine and then cooling it in the radiator.*
*The area of a circle can be found by measuring the radius and then multiplying its square by $\pi$*
*Impurities are eliminated by passing the liquid through a funnel containing a filter-paper.*

☐ **Exercise 9** Write more detailed explanations of how five of the following can be done. Use *by* + verb-*ing*. If necessary use more than one sentence for each answer. (You may also include diagrams if you wish.)

1 The circumference of a circle can be found by calculation.
2 Drinking water can be prepared from salt water by distillation.
3 A solid can be purified by crystallisation.
4 Atmospheric pressure can be found by Boyle's law.
5 Oven temperatures are usually controlled by thermostats.
6 Objects of different weight can be separated by a centrifuge.
7 The oxygen balance in the atmosphere is maintained by photosynthesis.
8 Rivers can be controlled by dams.
9 Heat can be obtained by fire.
10 In many parts of industrial countries the environment is threatened by pollution.

*By* is used with agents. It introduces *what* (or *who*) something is done by or the method by which it is done. However, English requires *with* and not *by* with instruments. *With* introduces the tool or instrument with which something is done. Therefore, only use *with* when you are definitely referring to an instrument (usually something you can hold in your hand.) Look at these examples:

*The door was opened by the teacher.*
*The door was opened by the wind.*
*The door was opened with a key.*
*The hole is made by a mechanic.*
*The hole is made by an animal.*
*The hole is made by the action of water.*
*The hole is made with a drill.*

o **Exercise 10** Complete these sentences putting in either *by* or *with*.

1 Holes can be made .... a drill.
2 Small diameters can be accurately measured ... a micrometer.
3 The heights of mountains are calculated .... surveyors.
4 The heights of mountains are calculated .... theodolites.
5 Wood is usually cut .... a saw.
6 Paint is usually applied .... a brush.
7 The metal of a bridge is usually protected .... paint.
8 Copper wires can be joined together .... using solder.
9 Copper wires can be joined together .... a soldering iron.
10 Quick but rather inaccurate calculations can be made .... a slide-rule.
11 A circle can be drawn .... a pair of compasses.
12 Written exercises are often corrected .... teachers .... red pens.
13 He will be examined .... the doctor.
14 If everything else fails, a door can be opened .... force!
15 These carpets are made .... hand.

## Must, should, and the passive

The modal passives *must be* verb + *ed* and *should be* verb + *ed* are commonly used to describe things which should or should not be done. This use of these modals was mentioned on page 36. The passive modals are particularly common in written instructions, warnings and notices:

*All library books should be returned to the library by the end of June.*

There are a number of useful verbs ending in *-en* which regularly occur in the passive with *must* or *should*. As the following table shows, these verbs are related to common adjectives and nouns.

| Adjectives | Nouns | Verbs |
|---|---|---|
| tight | | tighten (*make tighter*) |
| loose | | loosen (*make* ....) |
| weak | | weaken (*make* ....) |
| deep | | deepen (....) |
| short | | shorten (....) |
| wide | | widen (....) |
| strong | strength | strengthen (....) |
| high | height | heighten (....) |
| long | length | lengthen (....) |

△ **Exercise 11** Write out and complete these sentences. Notice the (;). Here is an example:

*The screws are too loose; .....*
*The screws are too loose; they should be tightened.*

1 The screws are too tight; ....
2 The bridge is too weak; ....
3 The road is too narrow; ....
4 The aerial is too short; ....
5 The harbour is too shallow; ....
6 The supports are lacking in strength; ....
7 The valve on this gas cylinder is not fixed firmly enough; ....
8 The entrance channel to the port is too restricted; ....
9 The radio signal from this transmitter has a range of only 120 kilometers; ....
10 With the recent introduction of larger lorries the road clearance of 3·5 meters under this bridge is unsatisfactory; ....

Instructions should be clear. (If they are not clear, there is no point in writing them.) Instructions can be made clearer by using words like *first, secondly, thirdly, then, next, finally.* These are organizing words. If they begin sentences, they should be followed by commas.

○ **Exercise 12** Here are two groups of sentences in the wrong order. (In other words, the instructions are not clear.) Give a title to each passage, and write out the passages putting the sentences in the right order. Use organizing words where necessary.

1 The letter should be put in the letter-box. Stamps of the correct value should be fixed at the top right-hand corner. The address should be clearly written underneath. The name of the person who is to receive the letter should be written in the middle of the envelope. A clean

envelope should be obtained. The address of the writer should be written on the back of the envelope.

2  When it is poured care should be taken not to breathe in the fumes. It should be kept in a cool, dark place. Nitric acid should be always handled with great care. The bottle should be immediately returned to its proper place. When it is being used it should be poured very carefully.

**Exercise 13**  Give written instructions for two of the following:

1  How to plot a graph
2  How to use a bunsen burner
3  How to mend a fuse
4  How to use a T-square
5  How to use balancing scales
6  How to find the center of an equilateral triangle
7  What to do when someone cuts himself badly
8  How to write instructions!

## Passives and infinitives

Here is a simple statement of fact:

*The world population is 3,500 million.*

In Unit 2 we saw that we can show that the number is not precisely known by writing:

*The world population is approximately 3,500 million.*

We can now express this uncertainty in a different way:

*The world population is thought to be 3,500 million.*
*The world population is believed to be 3,500 million.*
*China is estimated to have a population of 750 million.*
*China is said to have a population of 750 million.*

In contrast, certainty can be expressed by *be + known + to*:

*China is known to have a population of over 700 million.*

Now notice the different meanings of *can* in:

*For the purposes of this problem, g can be assumed to be 960 cm per sec$^2$.*
*If this experiment is carefully carried out, g can be shown to be 960 cm per sec$^2$.*

Finally, notice that *be + said + to* has a second meaning:

*If a body increases in speed, it 'accelerates.'*
*If a body increases in speed, it is said to 'accelerate.'*

○ **Exercise 14** Rewrite these simple statements using one of the passive forms below:

$$(\text{modal}) + be \begin{cases} thought \\ believed \\ estimated \\ said \\ known \\ assumed \\ shown \end{cases} to + \text{verb}$$

1  Africa has a population of 250 million.
2  If no force acts on a body it is 'in equilibrium.'
3  The moon has several dusty areas.
4  This liquid is poisonous.
5  The world has mineral oil reserves for another hundred years.
6  The opposite angles of a parallelogram are equal.
7  The shade temperature in the Sahara reaches 55° C.
8  A fluid which resists relative motion within itself is 'viscous.'
9  The new physics laboratories will cost $120,000.
10  A large nuclear power station runs on only 50 kg of uranium a year.

The previous exercise practised passives followed by infinitives. The passive infinitive itself is used in a number of different situations. One is after the modal *has/have to*:

$$The\ experiment \begin{cases} must \\ should \\ has\ to \\ will\ have\ to \end{cases} be\ repeated.$$

The infinitive is also used after *be* in a special structure. Consider these pairs of sentences:

(a) *The company will open the new factory next month.*
   *The new factory will be opened next month.*
(b) *The company is to open the new factory next month.*
   *The new factory is to be opened next month.*

(a) makes a prediction; the writer predicts that the factory.....
(b) suggests that those who know have stated or announced that the factory.....

It is not surprising, therefore, that *be* + infinitive is particularly used for reporting news. Here are some further examples:

*The Minister of Agriculture is to attend the next F.A.O. meeting.*
*The project is to be discussed by the committee.*
*All children are to be vaccinated.*
*The chemistry laboratory is to be extended because of the great increase in the number of students.*

△ **Exercise 15** Rewrite this paragraph, completing it by using the verbs underneath. Each verb should only be used once.

The plans for the new road .... in the near future. However, certain facts ....  . An entirely new route .... and the distance between the two cities .... by nine kilometers. It is possible, therefore, that a number of existing houses ....  . The road .... to carry up to 2,000 vehicles an hour and it .... a width of 12 meters. The total cost .... to exceed £5 million. The plans .... to state that only local building materials ....  .

| | |
|---|---|
| *are to be announced* | *will have* |
| *is designed* | *are already known* |
| *will have to be destroyed* | *will be reduced* |
| *is estimated* | *will be taken* |
| *are expected* | *are to be used* |

## Passive and active

The last few pages have dealt with the passive in scientific English. However, you should not get the idea from these pages that you must use a passive whenever possible. All science writers move from active to passive and back again according to what kind of statement they want to make. Indeed, it is possible to have too many passives. Here is a sentence with too many passive verbs:

*Students are informed that the substances to be analysed are to be labelled A, B, C and should be handled with care.*

○ **Exercise 16** Read the following passage and cross out the wrong alternatives.

Potential energy *uses/is used* to refer to the energy which a body *contains/is contained* because of its position. For example, a weight hanging above the ground *possesses/is possessed* potential energy. When a body *lifts/is lifted*, work *must do/must be done*. Some of this work *wastes/is wasted*, but some of it *may store/may be stored* inside the body as potential energy. When the body *releases/is released*, this stored energy *sets free/is set free* and some of it *may use/may be used* for doing work. The raised weight, for instance, *can use/can be used* to run a clock.

☐ **Exercise 17** Study the figure below and write an explanation of the cycle of a four-stroke petrol engine. Use either the active or passive. (a) has been done for you, 'notes' have been given for (b), and three words have been listed for (c).

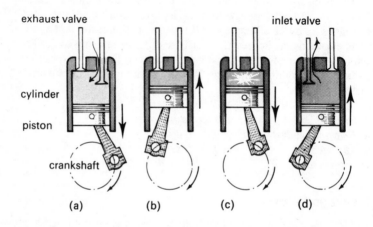

exhaust valve

inlet valve

cylinder

piston

crankshaft

(a)   (b)   (c)   (d)

(a) The exhaust valve closes and the inlet valve opens. The piston descends and creates a vacuum. As a result, a mixture of petrol and air is drawn in through the inlet valve.

(b) both valves closed—piston up cylinder—mixture compressed by piston

(c) spark plug, ignite, combustion

△ **Exercise 18** Study these pairs of grammatically correct sentences. Decide whether (a) the active sentence is better, (b) the passive sentence is better, or (c) both sentences are equally good. Then try and explain why.

1(a) You often find oil in limestone formations.
 (b) Oil is often found in limestone formations.
2(a) White surfaces reflect sunlight.

(b) Sunlight is reflected by white surfaces.

3(a) Factories produce millions of radios each year in Japan.

(b) Millions of radios are produced each year in Japan.

4(a) All wires resist the passing of an electric current.

(b) The passing of an electric current is resisted by all wires.

5(a) People largely base modern civilisation on iron and steel.

(b) Modern civilisation is largely based on iron and steel.

# Unit 5 More informative statements-relative clauses

## Introduction

So far we have dealt with sentences which contain only one finite verb. It is now time to consider sentences with two finite verbs. The most common two-verb sentences in scientific English are sentences containing relative clauses. Consider these four pairs of simple sentences:

(a) *An object is left in the sun. It becomes hot.* (Passive sentence + active sentence)
(b) *Wires are insulated with a plastic covering. They are used to carry electrical current.* (Passive + ....).
(c) *A plane seats 300 passengers. It is already in service.* (.... + ....)
(d) *The force holds the solar system together. The force is called gravitation.* (.... + ....)

These pairs of sentences can be made into single sentences by using relative clauses (the relative clauses are in italics.):

(a) An object *which is left in the sun* becomes hot.
(b) Wires *which are insulated with a plastic covering* are used to carry electric current.
(c) A plane *which seats 300 passengers* is already in service.
(d) The force *which holds the solar system together* is called gravitation.

What are the four main clauses?

(a) .....
(b) .....
(c) .....
(d) .....

The most important points about relative clauses are:

(a) Every written English sentence must have a main clause and a relative clause is never a main clause.

(b) Relative clauses begin with a *wh*-word. *Which* is by far the most common *wh*-word in scientific English. *Who* is used for people.

Relative clauses are used to avoid writing a series of very short sentences. They also enable a writer to keep the most important information for the main clause and to use a relative clause for the less important information. The main clause generally describes the result or demonstrates the principle. The relative clause generally describes the conditions and circumstances. Study these two examples:

(conditions/circumstances)    *An object is left in the sun.*
(result/principle)             *An object becomes hot.*

*An object which is left in the sun becomes hot.*

result/principle           *Iron can be shaped.*
conditions/circumstances    *Iron becomes red-hot.*

*Iron which becomes red-hot can be shaped.*

(When you do Exercise 1 keep the distinction between results and conditions in mind.)

## Passive relative clauses

**Exercise 1** Join the following pairs of sentences by using relative clauses. Notice that passive relative clauses must contain some part of *be*. The first one has already been done.

1 An object is left in the sun. It becomes hot.
2 A bottle is dropped on a stone floor. It usually breaks into pieces.
3 A balloon is filled with a gas lighter than air. It rises off the ground.
4 A disease is caused by a virus. It is often difficult to cure.
5 Paper is made from alfalfa. It is expensive.
6 Water is taken from a river. It is rarely pure enough to drink.
7 Pieces of iron are left in the rain. They become rusty.
8 An engine is run at maximum speed for a long time. It may start to overheat
9 Roads are surfaced with tarmac. They are faster than roads with surfaces of stones.
10 A wire is made of platinum. It costs more than a silver wire.

**Exercise 2** Join pairs of sentences as in the previous exercise. This time, however, the sentences have been mixed up. (Always choose one sentence from the left-hand column and one from the right-hand column.)

| | |
|---|---|
| Pieces of iron are left in the rain. | They should be read carefully. |
| The work is taken out of a system. | It solidifies. |
| A gas is subjected to very high pressure. | They usually contain a bi-metallic strip. |
| The instructions are written on the box. | They are used in the canning industry. |
| The thermometers are used to measure high temperatures. | It is less subject to corrosion. |
| The rocks are found at a depth of more than 20 miles under the earth's surface. | They soon begin to oxidize. |
| Steel is coated with paint. | They are semi-solid. |
| Potassium is found as a mineral. | They are called electrodes. |
| Sheets of iron are plated with tin. | It can never be greater than the work put in. |
| The two plates are immersed in the electrolyte. | It can be used as a fertilizer without chemical treatment. |

In the previous two exercises the relative clauses have followed nouns that are subjects of main clauses. (This is another instance of 'fronting.') However, relative clauses can also follow nouns in other positions:

| | |
|---|---|
| after nouns as complements | *A saw is a tool which is used for cutting wood.* |
| after nouns as objects | *The figure shows an apparatus which can be used to measure the specific heat of a metal.* |
| after a prepositional phrase | *The information is stored in a computer which is situated in a separate building.* |

☐ **Exercise 3** Read these sentences and then carry out the instructions.

1 Information about the weather comes from weather observations stations.
2 They are found in every part of the world.
3 There is, for example, a meteorological office at the South Pole.
4 It is supplied by aeroplane.
5 Information is also obtained from weather ships.
6 They are usually anchored in certain fixed places.
7 Meteorological observations are made at every important airport.
8 From the airport information about local weather conditions is sent to incoming planes.

9  Dangerous storms are sometimes carefully tracked.
10  Aircraft are often used for this.
11  Recently, more general information has been transmitted by weather satellites.
12  These are fixed in orbits.
13  These keep them stationary in relation to the earth's surface.

Choose a suitable title and rewrite the sentences as a passage of continuous English.

Join these sentences using relative clauses: 1 and 2; 3 and 4; 5 and 6; 7 and 8; 9 and 10; 11, 12, and 13. (7 and 8—*from which*; 9 and 10—*for which*.)

The relative clauses in 7 and 8, and 9 and 10 begin with prepositions. This preposition + *wh*-word structure presents difficulties to many learners of English. Here is another example:

(a) *Retorts are made from glass. The glass must be fire-proof.*

(b) *The glass (retorts are made from glass) must be fire-proof.*

(c) *The glass which retorts are made from must be fire-proof.*

(d) *The glass from which retorts are made must be fire-proof.*

Sentences (c) and (d) are both correct. However, (d) is more common in written scientific and technical English.

Here is a further example:

*Laboratories are rooms. Experiments are conducted in them.*

*Laboratories are rooms which experiments are conducted in.*

*Laboratories are rooms in which experiments are conducted.*

**Exercise 4**  In these sentences only one of the choices is right. Cross out the wrong *wh*-phrases.

1  The glass *which/from which/by which* retorts are made must be fire-proof.

2  A rectifier is a device *which/by which/in which* allows an electric current to flow in only one direction.

3  There are several ways *for which/in which/of which* sunshine can be controlled to provide heat.

4  A container of hot water gives out heat to the objects *which/by which/in which* it is surrounded.

E

5 Effort and load are forces *whom/through which/which* can be measured experimentally.

6 Acoustics is a branch of physics *in which/which/for which* the properties of sounds are studied.

7 A drill is an instrument *on which/round which/with which* holes are made.

8 A proton is a positive particle *which/by which/of which* forms part of the nucleus of an atom.

9 The process *for which/by which/which* plants build up glucose is called photosynthesis.

10 A turbine is an engine *round which/in which/by which* blades are forced to rotate.

Now notice these changes:

(a) *In which/At which* can become *where* when it clearly refers to a place:

*Airports are places at which special attention is paid to the weather.*
*Airports are places where .....*

*Warehouses are places in which goods are stored.*
*Warehouses are places where .....*

*The temperature at which water boils depends on the pressure.*
Not *The temperature where ....* (why?)

(b) *By means of which* can become *whereby*:

*A fuse-box is a device by means of which excessive loads are avoided.*
*A fuse-box is a device whereby excessive loads are avoided.*

(c) *Of which* can become *whose*:

*Venus is a planet of which the surface temperature is thought to be at least 200° C.*

This is the normal way in which prepositional relatives are formed. However, with *of which* it is more usual to find either:

*Venus is a planet the surface temperature of which is thought .....*
or:
*Venus is a planet whose surface temperature .....*

(Notice that *whose* is not restricted to references to people.)

**Exercise 5** Complete as least ten of the following, using passive relative clauses.

1 A piece of iron which .... becomes rusty.
2 Roads which .... are slower than roads surfaced with tarmac.
3 Water which .... is usually pure enough to drink.
4 Libraries are places where .....
5 The cloth from which .... should be water-proof.
6 Substances which .... are called adhesives.
7 Pressure is measured by dividing the force by the area over which .....
8 The center of gravity of a body is the point where .....
9 An object which .... displaces a volume of water equal to its own volume.
10 The solar energy which .... can be released by burning.
11 All the materials in a building must be strong enough to carry the loads which .....
12 A medical thermometer contains a constriction through which .....
13 A graduated tube with a tap for measuring the volume of liquid which .... is known as a burette.
14 The bomb which .... was equal to 20,000 tons of T.N.T.
15 Efficiency is the ratio between the energy which .... and the energy which .....

## Active relative clauses

Like passive clauses, active relative clauses in scientific English usually begin with *which*. They can also begin with *who*, *where*, *whose*, *that*, or a preposition + relative. Here are some examples:

*The Moto 1100 is a small family car. It has seats for 4 people.*
*The Moto 1100 is a small family car which has seats for 4 people.*

*The man teaches physics. He is a graduate of Delhi University.*
*The man who teaches physics is a graduate of Delhi University.*

*The room has only one small light-bulb. He works in the room.*
*The room in which he works has only one small light-bulb.*

**Exercise 6** Join these pairs of sentences using active relative clauses.

1 A plane seats 300 passengers. It is already in service.
2 Students operate lathes. They must observe the safety rules.
3 Steel contains very little carbon. It is known as mild steel.
4 A hydro-electric power station produces 4½ million kw. It is in operation at Bratsk in Russia.

5 A meteorite weighs 59 tons. It was discovered in South Africa in 1920.

6 The contractors are building the new Science Block. They hope to finish by December.

7 A non-metal resembles carbon in its physical properties. The non-metal is silicon.

8 Tungsten is a heavy metal. It melts at over 3,000° C.

9 Students do not check their calculations. They often make mistakes.

10 Some of the tankers will have a displacement of nearly half a million tons. Japanese companies are now building them.

△ **Exercise 7** Rewrite these sentences completing them with active relative clauses. The first one has already been done.

1 The man .... is a graduate of Delhi University.

2 A geometrical figure .... is called a polygon.

3 A body .... is said to be accelerating.

4 Radio telescopes pick up signals .....

5 A current of electricity .... produces heat.

6 Houses .... are rarely found in countries where snow is common.

7 A lorry is a vehicle .....

8 Carbon is an element .....

9 Stars .... can be seen with a telescope.

10 The temperature .... water .... depends on the atmospheric pressure.

## Reduced relative clauses

Many passive relative clauses can be reduced or shortened. In the reduced clause both the *wh*-word and *be* are omitted. Here are two examples:

*Pieces of iron which are left in the rain become rusty.*
*Pieces of iron left in the rain become rusty.*

*He uses an instrument which is called a spectroscope.*
*He uses an instrument called a spectroscope.*

Remember that there are only two correct structures:

*Pieces of iron which are left in the rain become rusty.*
*Pieces of iron left in the rain become rusty.*

These sentences are wrong:

*Pieces of iron which left in the rain become rusty.*
*Pieces of iron are left in the rain become rusty.*

○ **Exercise 8** Join these pairs of sentences using reduced relative clauses.

1 An object is left in the sun. It becomes hot.
2 A bottle is dropped on a stone floor. It usually breaks.
3 An exercise is written in a hurry. It may contain silly mistakes.
4 Diseases are caused by viruses. They are often difficult to cure.
5 Paper is made from alfalfa. It is expensive.
6 A wire is made of platinum. It costs more than a silver wire.
7 The technique is called seismic exploration. It is the first step in the discovery of oil.
8 The plan was prepared by the U.N. It was accepted by the government.
9 The cement was tested in the lab. It failed to set at high temperatures.
10 The new equipment was ordered by the college. It cost £13,500.

We use a reduced passive relative clauses because they are shorter and because a series of uninformative *which is* or *which are* structures can be avoided.

However, not all passive relative clauses can be reduced. Firstly, there are no reduced forms of relative clauses beginning with a preposition:

*The glass from which retorts are made must be fire-proof*

has no short form:

*The glass from retorts are made must be fire-proof* is wrong.

Secondly, there are normally no reduced forms of clauses containing modal verbs. The reason for this is obvious. If modal verbs are used to give extra meaning to the verb, then this extra meaning will be lost again in a reduced clause. Consider:

(a) *Here is a list of the experiments which will be done this year.*
(b) *Here is a list of the experiments which should be done this year.*
(c) *Here is a list of the experiments which have been done this year.*

Do (a) and (b) mean the same as:

*Here is a list of the experiments done this year?*

Does (c) mean the same?

It is important to distinguish very clearly between:

*Pieces of iron left in the rain become rusty* (where *left* is not a main verb, but a past participle in a reduced relative clause) and:

*He left pieces of iron in the rain* (where *left* is a main verb, being the past simple active of *to leave*.)

Therefore *Pieces of iron left in the rain* is not a complete sentence.

△ **Exercise 9** Decide whether the following are complete sentences. If you think they are complete, leave them. If they are not complete (i.e., reduced relative clauses) rewrite them, adding suitable words of your own.

  1  Pieces of iron left in the rain
  2  The bridge supported a weight of 250 tons
  3  A beam supported at both ends
  4  The sample weighed on the scales
  5  The man weighed it on the scales
  6  The bulldozer used 45 liters of petrol per hour
  7  The bulldozer used by the contractor
  8  Balloons filled with gas
  9  Fire damaged the buildings
 10  The buildings damaged by fire
 11  Light reflected in a mirror
 12  The light passed through a prism
 13  The analysis carried out in the lab
 14  He carried out the analysis in the lab
 15  The analysis he carried out in the lab

The active relative clause can also be reduced. In this case the *wh*-word (*which*, etc.) is omitted and the verb is changed to a present participle (verb + *-ing*). Here is an example:

*The man who lectures on Thursdays is an expert in dynamics.*
*The man lecturing on Thursdays is an expert in dynamics.*

Generally the present participle is formed by adding *-ing* to the base form of the verb. However, notice these spelling changes:

Verbs ending on consonant + *e* drop the *e*.

*lecture*    *lecturing*
*write*      *writing*
*live*       .....

Verbs ending in single vowel + single consonant double the consonant under the same conditions as for the past participle (see page 38).

| | | | |
|---|---|---|---|
| *order* | *ordering* | *stop* | ..... |
| *refer* | *referring* | *answer* | ..... |

**Exercise 10** Write out the present participles of these verbs:

| | | | |
|---|---|---|---|
| 1 study | 6 heat | 11 order | 16 refer |
| 2 stop | 7 prefer | 12 allow | 17 try |
| 3 use | 8 stay | 13 permit | 18 hope |
| 4 begin | 9 put | 14 develop | 19 watch |
| 5 design | 10 lose | 15 forget | 20 travel |

Again there are no reduced forms of active relative clauses which contain modals. Reduced forms also cannot be used when a noun or pronoun comes between the *wh*-word and the verb. Consequently

*This is the pen which he uses*

cannot reduce to

*This is the pen he using.*

**Exercise 11** Study the active relative clauses in these sentences. You will see that some of them can be reduced and some not. Rewrite those which can be reduced in the reduced form. The first one has already been done.

1 The man who lectures on Thursdays is an expert in dynamics.
2 The force that holds the solar system together is called gravitation.
3 Libraries are places where people study.
4 It is estimated that the number of people who live in China exceeds 700 million.
5 Fermentation is a chemical change which occurs when yeast is added to organic substances.
6 In many parts of the Sahara there is water which lies under the ground.
7 Leclanché cells are widely used in instruments which require an intermittent current.
8 An axis is an imaginary line about which a body rotates.
9 Some of the largest libraries have collections of books which may exceed a million volumes.
10 An Angstrom unit is a unit of length which equals 1/10,000 of a micron.

11 This improved type of light-bulb has a filament which should last at least 300 hours.
12 Mathematics is an essential subject for students who specialise in physics.
13 A theodolite is an instrument which surveyors use for measuring horizontal and vertical angles.
14 Rays which pass through a lens either converge or diverge.
15 Concrete which is strong in compression is likely to be highly weather-resistant.

○ **Exercise 12** Cross out the wrong alternative.

1 The test-tube (contained/containing) the solution was broken.
2 The cables (supplied/supplying) the electricity are out of order.
3 This report contains the results (taken/taking) during the test.
4 The temperatures (shown/showing) on the graph are given in degrees centigrade.
5 The techniques (used/using) in oil refining are complicated.
6 The liquid (flowed/flowing) through the pipe is low-grade oil.
7 The information (received/receiving) by the committee was passed on to the Minister.
8 Aluminium is a metal (produced/producing) from bauxite.
9 A new synthetic material (developed/developing) last year is already being produced on a commercial scale.
10 The metal (surrounded/surrounding) the engine must be able to withstand a temperature of 3,000° centigrade.

△ **Exercise 13(a)** Rewrite the passage below completing it with relative clauses (active or passive, reduced or not reduced.) Use this Table

**Verb-forms from five chapters of a physics test (in per cents)**

| chapters | 1 | 2 | 3 | 4 | 5 |
|---|---|---|---|---|---|
| Active | 40 | 36 | 50 | 34 | 46 |
| Passive | 36 | 32 | 18 | 40 | 24 |
| *Be* as a main verb | 24 | 32 | 32 | 26 | 30 |

The information .... describes principally the frequency of active and passive in five different chapters .... . It will be seen that the chapter .... is Chapter 4 and, in fact, this chapter is the only one .... . However, the percentage differences between active and passive are generally rather small and the one chapter .... is the third. Particularly consistent figures were obtained for the percentage occurrence of the main verb *be*, .... .

□ **Exercise 13(b)** Write up the following notes into passages of continuous English. Here is an example:

*balloon filled gas lighter than air—rises off ground*
*A balloon which is filled with a gas lighter than air rises off the ground.*

1 information about weather comes from 1000s obs. stations—found in all parts of world: this info. sent to main centers—where co-ordinated: in main centers exist computers—used—analyse data: these computers give forecasts—predict weather several days in advance.

2 amount of heat—emitted by wire—depends on resistance—metal offers to current: nickel chrome alloys—high R.—used in heating coils of elec. fires: *Ni Cr* expensive alloy—80% nickel & 20% chrome: 2 further properties—without them couldn't be used—high melting point & does not corrode.

# Unit 6  Definitions

## General definitions and a definition formula

Definitions occur frequently in many types of scientific writing because it is often necessary to define certain operations, substances, objects or machines. But what is a good definition?

Firstly, a definition is not an example. It is not intelligent to try and define something by giving an example of it. An example may follow a definition, but it may not take its place.

*Conduction is when you hold a glass in your hand and it becomes warmer.* (Is this a good definition? If not—why not?)
*A scientific law is a law like Ohm's law.* (Is this a good definition? If not—why not?)

Secondly, the first part of a definition should be general. The details should be left until later.

*A school is a place where you find blackboards.* (Is this a good definition? If not—why not?)
*A man is an animal with eight fingers and two thumbs.* (Is this a good definition? If not—why not?)

This means that the thing to be defined should be described first in terms of its general class, then in terms of its particular properties, qualities, uses or origins. This could be expressed as:

$$T = G + (d_a + d_b + d_c + \ldots\ldots d_n)$$

where $T$ equals the thing to be defined
where = equals *be*
where $G$ equals a general class word
where $d_a, d_b$, etc. are the properties which distinguish $T$ from the other members of the general class.
Here is an example:

*A catalyst (T) is a substance (G) which alters the rate at which a chemical reaction occurs ($d_a$), but is itself unchanged at the end of the reaction ($d_b$).*

i.e., $T = G + d_a + d_b$

Now look at these two lists of words. The first list contains general class words; the second does not.

## General class words

| a fruit | a metal | a machine |
| a gas | a liquid | a measure |
| a container | a figure | a science |

## Examples

| iron | an orange | a banana | copper | hydrogen |
| a triangle | a diesel engine | siver | nitrogen | a test-tube |
| ink | a lathe | a box | a quadrilateral | botany |
| geology | a liter | biology | water | an inch |
| an apple | oxygen | a crane | oil | a centimeter |
| a circle | a bottle | | | |

**Exercise 1** Make nine sentences of this form:

*The following are fruits: an apple, an orange, a banana.*

Choose a general class word and three examples each time. The first one has already been done.

Although there are several possible ways of writing definitions in English, there is one way which is much commoner than the others. It will be called the definition formula. Here it is in its simplest form:

An $\left\{ \begin{array}{c} x \\ y \end{array} \right\}$ is a/an .... general class word + *wh*-word .....

where $x$ is a countable noun
where $y$ is an uncountable noun
Here are some examples:

| An $x/y$ | is | a/an .... | class word | *wh*-word ..... |
|---|---|---|---|---|
| *A school* | *is* | *a* | *place* | *where children learn to read and write.* |
| *An engineer* | *is* | *a* | *person* | *who designs machines, buildings or public works.* |
| *Aluminium* | *is* | *a* | *metal* | *which is produced from bauxite.* |
| *Aluminium* | *is* | *a white* | *metal* | *which is light, weak, and resistant to corrosion.* |

Notice that *the* is not used with the subject because definitions are general statements. Also notice that in all the examples so far the main verb has been *is*. Other forms of *be* are not common. Other main verbs are also uncommon, although *can be defined as* is sometimes used.

△ **Exercise 2** In the following sentences the ten definitions have been mixed up. Write out the ten definitions correctly.

| An x/y | is | a/an | class word | wh-word | ..... |
|---|---|---|---|---|---|
| A plane | is | an | person | which | is used for cutting things. |
| A dentist | is | a | animal | which | consists of only one cell. |
| A dynamo | is | an | place | who | attracts bodies towards the center of the earth. |
| A triangle | is | a | machine | where | measures temperature. |
| Gravity | is | a | instrument | who | has three sides. |
| A shop | is | a | figure | which | generates electricity. |
| A thermometer | is | a | force | which | takes care of people's teeth. |
| A biologist | is | an | machine | which | things are bought and sold. |
| A knife | is | a | person | which | studies living organisms. |
| An amoeba | is | a | instrument | which | flies through the air. |

The word *device* is commonly used in writing definitions. This is because *device* is a kind of "general" general class word that can be used to refer to anything that has been invented or constructed. Look at these examples:

*A machine is a device which enables us to use forces more conveniently.*
*An engine is a device which converts one form of energy into another.*
*A geiger counter is a device which measures radiation.*
*A chronometer is a device which measures time.*
*A barometer is a device which measures atmospheric pressure.*

Notice how a *device* can be broken down into these four general class words:

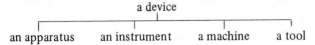

a device

an apparatus    an instrument    a machine    a tool

It is not always possible to describe exactly the difference between these four general class words. However, they are usually used as follows:

*An apparatus* is a number of devices which are put together for a particular purpose, as in physics or chemistry experiments.
*An instrument* is a device which is used in doing something, often of a sensitive nature. Typical examples are a microscope and an ammeter.
*A machine* is a mechanical device which is used to provide power.
*A tool* is a simple device, often without any moving parts. Examples are a hammer and a spanner.

○ **Exercise 3** Here are definitions of ten devices. Cross out the wrong general class words.

1 A tape-recorder is a tool/machine which records sound.
2 A Wheatstone bridge is a machine/apparatus which measures electrical resistance.
3 A screwdriver is a tool/apparatus which tightens or loosens screws.
4 A drill is an instrument/apparatus which bores holes.
5 A condenser is a machine/tool which converts vapour into liquid.
6 An ammeter is a machine/instrument which measures electrical current.
7 A fan is an instrument/apparatus which circulates air.
8 A saw is an instrument/tool which cuts wood.
9 A clock is an instrument/tool which measures time.
10 A generator is an apparatus/machine which produces electricity.

Definitions are often completed by a passive relative clause. Here are two examples from previous pages:

*Aluminium is a metal which is produced from bauxite.*
*A knife is an instrument which is used for cutting things.*

Here are two more examples:

*An alloy is a metallic substance which is composed of two or more elements.*
*A pump is a machine which is used for transferring a liquid or a gas from one place to another.*

69

You will notice that two of the examples contain *which is used for*. For obvious reasons this phrase is very common in definitions.

△ **Exercise 4** Write definitions of ten of the following using *which is used for* as part of the definition completion.

| | | |
|---|---|---|
| 1 a knife | 2 a slide-rule | 3 a thermometer |
| 4 a pen | 5 a tractor | 6 a micrometer |
| 7 a drill | 8 a spanner | 9 a speedometer |
| 10 an axe | 11 a crane | 12 an ammeter |
| 13 a ruler | 14 a telescope | 15 a barometer |

Look again at this example:

*Aluminium is a metal which is produced from bauxite.*

You will remember that it is also possible to use a reduced relative clause and write:

*Aluminium is a metal produced from bauxite.*

△ **Exercise 5** Read carefully the definition completions given below. (Remember that they are all reduced passive relative clauses.) Try and discover the things which are being defined and write out as many whole definitions as you can. The first has been done.

1 .... produced from bauxite.
2 .... used for boring holes.
3 .... composed of two or more elements, one of which must be a metal.
4 .... used for detecting very small electric currents.
5 .... extracted from its ore in a blast furnace.
6 .... represented by the symbol *Pb*.
7 .... used for separating solids from liquids.
8 .... prepared in the laboratory by the action of dilute sulphuric acid on zinc.
9 .... used for measuring the distance between two points on a curved surface.
10 .... composed of cement, sand, small stones, and water.
11 .... represented by the symbol *Ca*.
12 .... obtained by multiplying a number by itself.
13 .... usually composed of about 70% copper and 30% zinc.
14 .... obtained as a by-product in the manufacture of coal-gas.
15 .... principally used as a conductor and in the cores of pencils.
16 .... represented by the symbol *Fe*.

17 .... prepared in the laboratory by the action of dilute hydrochloric acid on marble chips.
18 .... derived from water after replacing one of its hydrogen atoms with some other atom.
19 .... generally plotted between two axes at right-angles to each other.
20 .... caused by the pull exerted by the sun and moon on the water of the sea.

**Exercise 6** Change the following 'notes' into written definitions. In addition to finding the correct form of the verb, you must also think about articles and prepositions. To help you, the verbs are in italics. Here is an example:

*knife—instrument—use—cut—things*
*A knife is an instrument used for cutting things.*

1 aluminium—metal—*produce*—electrolysis
2 dynamics—branch—mathematics—*concern*—moving bodies
3 pressure—*define*—force—unit—area
4 concrete—substance—*make*—mixture—cement, gravel, water
5 volume—measure—amount—space—*occupy*—body
6 force—gravity—attraction—*cause*—earth's central mass
7 pylon—tall steel structure—*use*—*carry*—high voltage wires
8 saw—instrument—*use*—*cut*—wood
9 plastics—synthetic polymers—*compose*—long chain-like molecules
10 specific heat—material—heat—*require*—*raise*—temperature—one gram—material—1° centigrade

We have already seen that reduced relative clauses are often used in completing definitions. A further reduction is possible, but only with the phrase *used for*:

*A knife is an instrument which is used for cutting.*
*A knife is an instrument used for cutting.*
*A knife is an instrument for cutting.*
*A thermometer is an instrument for measuring temperature.*
*A thermostat is a device for regulating temperature.*

(Notice that the *for* structure can only be used with devices.)

**Exercise 7** Write definitions of ten of the following, using the formula: An *x/y* is a .... (class-word) for .....

| | | |
|---|---|---|
| 1 a knife | 2 a pair of scissors | 3 a pipette |
| 4 a ruler | 5 a funnel | 6 a voltmeter |

|   |   |   |
|---|---|---|
| 7 a crane | 8 a T-square | 9 a barometer |
| 10 a test-tube | 11 an altimeter | 12 a carburettor |
| 13 a microphone | 14 an oscillascope | 15 a scalpel |

Of course the passive is not always used in scientific definitions. You will meet many definitions containing the third person singular of the Present Simple active. (Look back to Exercise 2; how many of the sentences have the second verb in the Present active?)

○ **Exercise 8** Join pairs of sentences by using relative clauses. Here is an example:

*Stainless steel is an alloy. It does not corrode.*
*Stainless steel is an alloy which does not corrode.*

However, the sentences below have been mixed up. Join one sentence from the left-hand column and one from the right-hand column.

| | |
|---|---|
| Stainless steel is an alloy. | It gives a magnified image of an object. |
| An insect is an animal. | It contains acetic acid. |
| A simple microscope is a device. | He has a university degree. |
| Tungsten is a metal. | It consists of two parts of hydrogen and one part of oxygen. |
| A spirit-level is an instrument. | It has a body divided into three parts. |
| A graduate is a man. | It contains minerals. |
| An ore is a natural substance. | It resists corrosion. |
| Water is a liquid. | It contains a large proportion of copper. |
| Vinegar is a liquid. | It retains hardness at red heat. |
| Brass is an alloy. | It works on the principle of an air-bubble. |

It is also possible to use verb + -*ing* instead of *wh*-word + verb + *s*. Here is an example of one of these reduced active clauses:

*A tangent is a straight line which touches a curve at one point.*
*A tangent is a straight line touching a curve at one point.*

This verb + -*ing* is particularly common with the verbs *contain* and *consist of*. Do not confuse the meanings of the two verbs:

*This box contains filter papers.*
*This box consists of four sides, a bottom, and a lid.*

**Exercise 9** Complete these sentences, using either *containing* or *consisting of*.

1 An ordinary thermometer .... mercury.
2 An ordinary thermometer .... a closed tube with a bulb at one end, a small amount of mercury, and a graduated scale.
3 Brass is an alloy .... a large proportion of copper.
4 A library is a room or a series of rooms .... a large number of books.
5 A pair of callipers is an instrument .... two curved pointers linked at one end.
6 An English sentence is a piece of language .... at least one main verb.
7 River water usually .... a number of impurities.
8 A pair of spectacles is a device for correcting eye-sight .... a lens for each eye and a frame.
9 Coal is a material .... carbon and various carbon compounds.
10 Coal is a complex material .... a large number of different elements.

So far the general class word has always been followed by a *wh*-word or by a reduced relative clause. However, there may be a preposition before the *wh*-word. This happens when the subjects of the two parts of the definition sentence are not the same, as in this example:

*Acoustics is a branch of physics. The properties of sounds are studied in it.*
*Acoustics is a branch of physics in which the properties of sounds are studied.*

**Exercise 10** Complete ten of these definitions.

1 A drill is an instrument with which .....
2 A library catalogue is a collection of cards on which .....
3 A pair of compasses is an instrument with which .....
4 A foundation is a base on which .....
5 A lathe is a machine on which .....
6 Geometry is a branch of mathematics in which .....
7 An axis is an imaginary line about which .....
8 A telescope is an instrument through which .....
9 An airport is a place at which .....
10 A filter is a device through which .....
11 An ion is an electrically charged group of atoms in which .....
12 A transformer is a device by means of which .....
13 A telephone is a device whereby .....
14 A convex lens is a curved piece of glass in which .....
15 Frequent use of the passive is a way of writing English in which .....

F

Finally, there are two ways of writing scientific definitions which do not make use of relative clauses at all. Look at these examples:

*A triangle is a plane figure which has three sides.*
*A triangle is a plane figure with three sides.*
*Tungsten is a metal which retains hardness at red-heat.*
*Tungsten is a metal with the property of retaining hardness at red-heat.*

○ **Exercise 11** Rewrite these sentences using either *with* or *with the property of.* (Read each definition carefully before deciding which to use.)

1 Carbon is an element which has an atomic weight of 12·01.
2 Caffeine is a substance which has a powerful action on the heart.
3 Cement is a powder which sets into a hard mass after it has been mixed with water.
4 A dye is a substance which can change the color of a meterial.
5 An electron is a particle with a mass of 9·107 x $10^{-28}$ grams.
6 An equilateral figure is a figure which has all sides equal in length.
7 Force is an outside agency which can change the state of motion or rest in a body.
8 Iron is an element which has an atomic weight of 55·85.
9 A pentagon is a plane figure which has five sides.
10 An insect is a small animal which has a body divided into three parts.

*Summary of the definition formula*

| | | |
|---|---|---|
| | *which is* verb + *-ed* ..... | (as in Ex. 4) |
| | verb + *-ed* ..... | (as in Ex. 5) |
| | *for* verb + *-ing* ..... | (as in Ex. 7) |
| *An x/y is a* class-word | *wh-*word verb + *s* ..... | (as in Ex. 8) |
| | verb + *-ing* ..... | (as in Ex. 9) |
| | preposition *wh-*word ..... | (as in Ex. 10) |
| | *with* noun-phrase ..... | (as in Ex. 11) |
| | *with the property of* verb + *-ing* | (as in Ex. 11) |

□ **Exercise 12** Write three definitions from each of the following groups:

1 energy, gravity, volume, heat, sound, friction, acceleration
2 chlorine, nitric acid, hydrogen, polyvinylchloride, carbon dioxide, combustion, oxidation, clouds

3 a cylinder, a pyramid, a sphere, a radius, a trapezium, an equation, a fraction
4 a dynamo, a diesel engine, a rheostat, a pulley, a beam, a prism, a derrick
5 a dictionary, a map, a definition, an index, a full-stop, an English sentence, a drawing board

## Specific definitions

Nearly all the definitions we have looked at so far have been general. In terms of language this means that the thing to be defined has usually been a single noun unaccompanied by other nouns or adjectives specifying it. Here is a simple general definition:

*A saw is an instrument used for cutting wood.*
... T.....= ............ G....... + ............... d...............

But now consider:

*A key-hole saw is a saw with a narrow blade,*
... T...........+ t   = ...t......+..........$d_a$..............

*used for cutting holes in wood.*
+...................$d_b$....................

This is a *specific* definition and *T* now equals the *type* of thing to be defined. In other words, we are now interested in defining a specific type of saw (the key-hole type) rather than defining a saw in general. Here is an example form an earlier exercise:

*an equilateral figure is a plane figure with all sides equal in length.*
...................+..........=.....................+..........................................

Notice that in the last two examples the nouns are repeated (*t* .. *t*).

   Now consider these two specific definitions:

*An equilateral triangle is a triangle with all three sides equal in length.*
*An equilateral triangle is a plane figure with all three sides ....*

The second of the sentences shows that a general class word can still be used. Therefore, the complete formula is:

$$T + t = \begin{Bmatrix} t \\ G \end{Bmatrix} + d_a + d_b + ... d_n$$

Indeed, sometimes a general class word (*G*) must be used. Compare:

*A suspension bridge is a bridge with a long central span suspended from cables.*
*A Wheatstone bridge is an apparatus for measuring the resistance of an electric circuit.*
(What is wrong with *A Wheatstone bridge is a bridge* ....?)

○ **Exercise 13** Complete these specific definitions, using the phrases given.

    (a) an alcohol thermometer, a Beckmann thermometer, a bimetallic thermometer, a gas thermometer, a maximum and minimum thermometer, a medical thermometer, a resistance thermometer

1 .... is a highly accurate thermometer which works on the principle that a small rise in temperature causes a comparatively large expansion in a volume of gas.
2 .... is a thermometer used to record the highest and lowest temperatures over a period of time.
3 .... is a narrow range thermometer containing a constriction so that body temperature can be read afterwards.
4 .... is a thermometer used principally for measuring temperatures below the freezing point of mercury.
5 .... is a thermometer which works on the principle that the electrical resistance of a conductor varies with temperature.
6 .... is a sensitive thermometer for measuring small changes in temperature up to 0·01 of a degree.
7 .... is an industrial thermometer consisting of a thin spiral of metal with one end fixed and the other attached to a pointer moving round a graduated scale.

    (b) a Daniell cell, a Leclanché cell, a photo-electric cell, a Voltaic cell

8 .... is an early type of cell consisting of cloth soaked in salt water placed between a copper disc and a zinc disc.
9 .... is a device used for detecting and measuring light.
10 .... is a primary cell containing a carbon rod surrounded by a mixture of magnanese dioxide and powdered carbon.
11 .... is a primary cell which contains a zinc rod standing in a porous pot containing dilute sulphuric acid.

    (c) high speed steel, medium steel, mild steel, stainless steel.

12  .... is an alloy containing about 18% tungsten and small percentages of chromium and vanadium.
13  .... is an alloy usually containing about 18% chrome and 8% nickel.
14  .... is an alloy containing about 0·25% carbon.
15  .... is an alloy with a carbon content of between 0·3 and 0·6%.

Finally, notice that in specific definitions it is not unusual to find that the noun after *is* has some form of qualification, as in:

*An equilateral figure is a plane figure .....*

How many examples of this are there in Exercise 13?

☐ **Exercise 14**  Write specific definitions of at least ten and not more than fifteen of the following:

| | |
|---|---|
| 1  a spring balance | 16  a weather balloon |
| 2  a filament lamp | 17  the periodic table |
| 3  an electric fire | 18  sedimentary rocks |
| 4  roof tiles | 19  potential energy |
| 5  an isosceles triangle | 20  inorganic compounds |
| 6  alternating current | 21  a tuning fork |
| 7  a horse-shoe magnet | 22  a diode valve |
| 8  a two-stroke engine | 23  a gramophone needle |
| 9  the Fahrenheit scale | 24  a lift pump |
| 10  seismic exploration | 25  a driving mirror |
| 11  a thermos flask | 26  the Plimsoll line |
| 12  an alarm clock | 27  radio waves |
| 13  a dovetail joint | 28  cumulus clouds |
| 14  integral calculus | 29  latent heat |
| 15  solid geometry | 30  inert gases |

## Expanded definitions

So far we have only considered one-sentence definitions. However, some of the sentences—especially in Exercise 13—are rather long. Long sentences are unnecessarily difficult for students to write. Every language teacher knows that the longer a sentence a student tries to write the more likely he is to make a lot of mistakes. Therefore, we now look at some ways of writing two-sentence definitions.

### Definition + example

The first exercise of this unit practised sentences of the form:

*The following are acids:— sulphuric acid, nitric acid, hydrochloric acid.*

Such sentences give examples. However, if the examples follow a definition, they are usually expressed in a different way. Here are two examples of this:

*An acid is a compound which neutralizes a solution of sodium hydroxide*
*Common examples are sulphuric and nitric acid.*
*An acid is a compound which neutralizes a solution of sodium hydroxide*
*Typical examples are sulphuric and nitric acid.*

Another way of following a definition with an example makes use of *such as*:

*An acid is a compound which neutralizes a solution of sodium hydroxide*
*such as sulphuric acid or nitric acid.*

Do not use *as* but *such as*. Notice the comma (,).

△ **Exercise 15(a)** Expand these definitions by adding examples. Use each of the two ways explained above at least twice.

1 A plane figure is a figure which has only two dimensions.
2 An equation is a mathematical expression in which the two sides balance.
3 An alkali is a substance which combines with acids to form salts.
4 Lenses are rounded pieces of glass used for controlling light rays.
5 A bridge is a structure which spans a gap.

△ **Exercise 15(b)** Write definitions plus examples of five of the following.

1 an alloy    5 a primary cell
2 an atom    6 a table
3 a tool      7 a knife
4 a fuel      8 liquids

*Definition + use*

An $x/y$ is a $Z$ ..... $\left.\begin{array}{l} \textit{Therefore,} \\ \textit{Consequently,} \\ \textit{As a result,} \end{array}\right\}$ $\left\{\begin{array}{l} \textit{it is used} ..... \\ \textit{one of its main uses} ..... \end{array}\right.$

*Aluminium is a metal which is light in weight.*

$\left\{\begin{array}{l} \textit{Therefore, it is used for making aircraft.} \\ \textit{Consequently, it is widely used for the manufacture of aircraft.} \\ \textit{As a result, it is widely used in the aircraft industry.} \\ \textit{Therefore, one of its main uses is in the manufacture of aircraft.} \end{array}\right.$

**Exercise 16** Write out and complete these definitions by expanding them as in the examples above. (The phrases in brackets may be of help.)

1 Aluminium is a metal which is light in weight.
2 Stainless steel is an alloy which resists corrosion.
3 Mercury is a liquid metal which has a high co-efficient of expansion.
4 Glass is a substance which has the property of being transparent.
5 Tungsten is a metal which retains hardness of red-heat. (Filaments in electric light-bulbs)
6 Copper is a metal which is a good conductor of electricity. (in the electrical industry)
7 Cement is a powder which sets into a hard mass after it has been mixed with water. (the building industry)
8 A dye is a substance which can change the colour of a material. (the manufacture of cloth) (the textile industry)
9 Gold is a precious metal which always shines brightly. (jewellery)
10 Phosphate is a mineral which is an essential element of high-yielding soil. (fertilizer)

*Definition + main parts*

An $x/y$ is a $Z$ ..... It consists of $\begin{cases} two \\ three \\ four \\ .... \\ these \end{cases}$ main parts: .....

*A pair of spectacles is a device for correcting eye-sight. It consists of three main parts: a lens for each eye and a frame.*

**Exercise 17** Write expanded definitions of five of the following:

1 a razor
2 a pair of scissors
3 a pair of compasses
4 a light-bulb
5 a bunsen-burner
6 a bicycle pump
7 a match-box
8 a slide-rule
9 a spring-balance
10 a cigarette lighter

## *Summary of expansions*

Definition formula +
{
*Common examples are a, b, c, and d.*
*Typical examples are a, b, c, and d.*
*Main types are a, b, c, and d.*
*such as a, b, c, or d.*
*Therefore, it is used .....*
*As a result, one of its main uses is .....*
*It consists of .... main parts: .....*
*Its main components are .....*
}

# Unit 7   Scientific statements referring to the past

The most typical kind of scientific writing is that used in scientific text-books. In such text-books only a small percentage of verbs are in either of the two main past tenses—the Past Simple and the Present Perfect. However, in certain other kinds of scientific writing the past tenses are much more frequent. These other types are principally:

histories of science and technology
some kinds of scientific and technical reports
scientific journalism (that is, news about science and scientists)

## Form of the Past Simple

The statement, negative, and question forms of the Past Simple can be seen below:

### Active

*The storm damaged the harbour last week.*
*They tested the new batteries yesterday.*
*They did not complete the work on time.*
*Did they finish the analysis yesterday?*

### Passive

*The harbour was damaged by the storm.*
*The new batteries were tested yesterday.*
*The work was not completed on time.*
*Was the analysis finished yesterday?*

Notice the similarity between the Present Simple and the Past Simple—they both require the 'empty' auxilary *do* in the active negative and question forms.

Here is a list of the most common irregular verbs used in scientific English.

| be | was | been | hold | held | held |
|---|---|---|---|---|---|
| bear | bore | borne | keep | kept | kept |
| become | became | become | know | knew | known |
| begin | began | begun | lead | led | led |
| bend | bent | bent | lean | lent | lent |
| bind | bound | bound | leave | left | left |
| break | broke | broken | let | let | let |
| bring | brought | brought | lie | lay | lain |
| build | built | built | light | lit/lighted | lit/lighted |
| burn | burnt | burnt | lose | lost | lost |
| choose | chose | chosen | make | made | made |
| come | came | come | mean | meant | meant |
| cost | cost | cost | meet | met | met |
| cut | cut | cut | put | put | put |
| deal | dealt | dealt | read | read | read |
| do | did | done | ring | rang | rung |
| draw | drew | drawn | rise | rose | risen |
| drive | drove | driven | run | ran | run |
| fall | fell | fallen | say | said | said |
| feed | fed | fed | see | saw | seen |
| feel | felt | felt | set(up) | set | set |
| find | found | found | shake | shook | shaken |
| fly | flew | flown | shut | shut | shut |
| forget | forgot | forgotten | spend | spent | spent |
| freeze | froze | frozen | spin | span | spun |
| get | got | got/gotten | stand | stood | stood |
| give | gave | given | strike | struck | struck |
| grow | grew | grown | swing | swung | swung |
| grind | ground | ground | tear | tore | torn |
| hang | hung | hung | think | thought | thought |
| have | had | had | understand | understood | understood |
| hit | hit | hit | wear | wore | worn |
| | | | write | wrote | written |

○ **Exercise 1** Rewrite, putting the verbs in brackets into the correct form (either active or passive). Here is an example:

*The crucible .... with a clamp. (hold)*
*The crucible was held with a clamp.*

1 The sun .... at 5.34 this morning. (rise)

2  The eclipse .... at 8.20 P.M. on 1st July, 1963. (see)
3  He .... the apparatus the day before the experiment. (set up)
4  The lesson .... by most of the class. (not understand)
5  The lenses for many eighteenth century optical instruments .... in Holland. (grind)
6  Early pumps .... by the wind. (drive)
7  The first planes .... at less than 100 km per hour. (fly)
8  Unfortunately, the measuring-rod .... in the previous experiment. (break)
9  Originally, the tower of Pisa .... at such an angle. (not lean)
10  This beam-balance .... a great deal of money. (cost)

## Use of the Past Simple

The Past Simple is normally used to describe actions which happened in the past and are now finished. Consider this statement:

*The first satellite, Sputnik 1, circled the earth 200 times.*

Because the Past Simple is used we know that the action is finished. We know that Sputnik 1 is no longer going round the earth.

Often, however, the Past Simple is used together with a time-phrase that refers to the completed past. (How many of the sentences in Exercise 1 contain such time-words and time-phrases?)

Notice particularly these ways of referring to the past:

*Early*
*The first*  } *generators produced only direct current.*

*Originally,*
*Initially,*  } *generators produced only direct current.*

These expressions are commonly used to refer to beginnings; in this case, the beginning stages of the development of .....

△ **Exercise 2** Complete these sentences. The words in brackets may be of help. Here is an example:

*Early boats .... (reeds)*
*Early boats were often built of reeds.*

1  Early generators .....
2  Early blast furnaces ..... (fuel—charcoal)
3  The first steam engine ..... (boiler pressure—only three atmospheres)
4  The first television sets ..... (small screens)

    5  Early internal combustion engines ..... (fuel–coal-gas)
    6  Early bridges .....
    7  Early methods of lifting water .....
    8  The first instruments for measuring temperature ..... (air)
    9  Initially, light bulbs ..... (filaments–platinum wire)
10  Originally, rubber ..... (Brazil)

As mentioned at the beginning of this unit, the Past Simple is usually used in writing technical reports.

△ **Exercise 3**  Here are the 'notes' of a simple technical report. Write them up into a continuous passage. (The report is fairly technical; is it better to use the active or passive?)

13/4/70  Royle and Brown collected samples of cement type 143 from World Oil/purpose–analysis of failure to solidify

14/4/70  R and B analyzed the composition of the cement/no useful results
They heated the cement to 200° C (prob. temp. at 8000′)/ nothing significant

15/4/70  They increased the pressure on the cement by 4/discovered weakness in type 143
Sent report to the Oil Company–report described conditions under which failed to set

It has already been said that the Present Simple is usually used when describing experiments. This is particularly true if standard or classical methods are employed. Consider this description:

## An experiment to measure atmospheric pressure (after Torricelli).

*First, a long glass tube is taken. The tube is closed at the top and is then completely filled with water. Next it is placed vertically in a large barrel half-full of water. When the bottom of the tube is opened, the water level in the tube only falls to a height of approximately 10 meters above the water level in the barrel. As a result, a vacuum is left in the upper part of the tube. The water in the tube is supported by the atmospheric pressure. The height of the column of water can therefore be used to measure atmospheric pressure.*

This is written in the Present Simple passive. But if we wished to describe how atmospheric pressure was originally measured by Torricelli, we would use the Past Simple active.

△ **Exercise 4** Complete the following passage:

## How Torricelli measured atmospheric pressure

The weight of air above our heads was first measured by the Italian scientist Torricelli in 1643. He took .....


It was said at the beginning of this unit that the Past Simple is also used in making statements about the history of science and technology.


△ **Exercise 5** Here are ten such statements. However, they have been mixed up. Write out the ten correct statements.

| | | | |
|---|---|---|---|
| The laws of gravity | was | invented | in the 27th century B.C.* |
| The Eiffel tower | were | discovered | in the 20th century A.D. |
| X-rays | was | invented | in the 15th century A.D. |
| The Roman alphabet | were | invented | in the 20th century A.D. |
| Television | were | discovered | in the 17th century A.D. |
| The Gizah pyramids | was | built | in the 20th century A.D. |
| The jet engine | was | built | in the 16th century B.C. |
| The Taj Mahal | was | discovered | in the 19th century A.D. |
| America | was | invented | in the 19th century A.D. |
| The transistor | was | built | in the 17th century A.D. |

Now consider:

*The first telescope was invented*
$\begin{cases} \text{in the seventeenth century.} \\ \text{in the early seventeenth century.} \\ \text{in 1609.} \\ \text{about 350 years ago.} \\ \text{over 350 years ago.} \end{cases}$

Notice these points:

(a) the *early* 17th century — the *middle of* the 17th century — the *late* 17th century
(b) *at* the beginning of, *in* the middle of, *at* the end of
(c) Statements of time can be 'qualified' in exactly the same way as statements of dimension by means of *approximately, over, under,* etc. (See Unit 2.)

* B.C. = Before Christ; A.D. = *Anno Domini*, a Latin expression meaning *Year of the Lord* and, hence, After Christ.

o **Exercise 6** Rewrite these sentences using 'qualified' time-statements. The first one has already been done.

  1 The first telescope was invented in 1609.
  2 A symbol for zero was probably invented in A.D. 595.
  3 Dynamite was invented in 1867.
  4 The first printing press in Europe was established in 1447.
  5 Man's first flight took place in 1783.
  6 The Suez canal was opened in 1869.
  7 Penicillin was discovered in 1928.
  8 Nitric acid was first distilled in 1250.
  9 D.D.T. was first used in 1940.
10 The first artificial satellite was put into orbit in 1957.

One of the few occasions when *by + who*-agent is used in scientific English occurs when making statements about inventors and discoverers:

*Dynamite was invented by Nobel.* (one person)
*Penicillin was discovered by Fleming and Florey.* (two people)
*The world's first television service was started by the B.B.C.* (a group of people, a company)
*The first satellite was put into orbit by the Russians.* (a people, a nation)

o **Exercise 7(a)** Complete at least five of the following:

  1 .... was invented by Edison
  2 .... was invented by Galileo.
  3 .... was discovered by Madame Curie.
  4 .... were discovered by Newton.
  5 .... was invented by Graham Bell.
  6 .... were built by the ancient Egyptians.
  7 .... was first developed by Bessemer.
  8 .... was designed by Sinan.
  9 .... were discovered by the ancient Greeks.
10 .... was first discovered by the Chinese.

△ **Exercise 7(b)** Complete five of the following, using either the active or the passive. Use different sentences to the ones you used in Exercise 7(a).

  1 .... discovered .....     6 .... analyzed .....
  2 .... invented .....      7 .... published .....
  3 .... built .....         8 .... determined .....
  4 .... made .....         9 .....developed .....
  5 .... organized .....    10 .... introduced .....

There is another common way of referring to inventors or discoverers; it takes two slightly different forms:

(a) *Euclid was the first person to write a systematic description of elementary geometry.*
(b) *The first person to write a systematic description of elementary geometry was Euclid.*

As these two examples suggest, this kind of sentence is often used when what has been invented or discovered cannot be expressed in two or three words. (Sentences like (b) may be preferred because they show 'fronting.')

△ **Exercise 8** Here are some 'notes.' Write them out as complete written statements like (b) above. Be careful about articles and prepositions. The first one has already been done.

1 first person—write—systematic description—elementary geometry—Euclid
2 first person—give—clear presentation—quadratic equations—Abu-jaafer
3 first person—compose—logarithm tables—Napier
4 first person—discover—angles—base—isosceles triangle—equal—probably—Thales
5 first person—prove—sum—angles—triangle—equal—two right angles—Pythagoras
6 first person—harden—rubber—sulphur—Charles Goodyear
7 first person—find—technique—converting pig iron—steel—Henry Bessemer
8 first book—contain—signs (+) and (−) —addition and subtraction—Widman's *Arithmatic*—publish—1489
9 first person ....—Newton
10 first person ....—Edison

## Form of the Present Perfect

The statement, negative and question forms of this tense are shown in this table.

### Active

*The water has already boiled for ten minutes.*
*The new microscopes have arrived.*
*The examinations have not begun yet.*
*Has the water boiled?*

Scientific statements referring to the past

**Passive**

*This has been known for many years.*
*All the acids have been identified.*
*Fermatt's last theorem has never been proved.*
*Has the analysis been completed?*

There are a number of adverbs of time that are commonly used with the Present Perfect. Notice that they take the same position as *not*:

He has $\left\{\begin{array}{l} not \\ never \\ recently \\ not\ yet* \\ still\ not \\ just \\ already \end{array}\right\}$ *published his results.*

His results have $\left\{\begin{array}{l} not \\ never \\ recently \\ already \\ not\ yet \\ still\ not \\ just \end{array}\right\}$ *been published.*

○ **Exercise 9** Rewrite these sentences, putting the verbs in brackets— sometimes verbs + adverbs—into the correct Present Perfect form. You must also choose the active or passive.

1 Medicine (make) great progress in the last twenty years.
2 Man (not yet discover) a cure for the common cold.
3 The whole area (already photograph) from the air.
4 Fortunately, the voltmeter (already mend).
5 The students (just complete) a long experiment.
6 The generator (break down) twice this week.
7 Chemists (develop) many synthetics during the last few years.
8 The examination results (just publish).
9 The World Health Organisation (supply) aid to many countries.
10 Since its invention nearly two hundred years ago, the steam engine (become) the most important single source of power that man (ever know).

* *Yet* can also come at the end of the sentence, and must do so in questions: *Has he published his results yet?*

88

## Use of the Present Perfect in descriptions

It was said in Unit 1 that the Present Simple is the tense used to describe scientific facts and scientific events. This remains generally true. However, there are a few occasions when it is either better or necessary to use the Present Perfect. Consider this short passage:

*A distillation flask is filled with a mixture of water and methylated spirit in a ratio of 2 to 1. The flask is then heated over a low flame. Distilling is continued until about 3 cm of the distillate has collected in the boiling-tube.*

There are .... verbs in the passage.
Three verbs are in the .... tense.
One verb is in the .... tense.
There is only one subordinate clause in the passage. This clause begins with the word ....; the verb in the clause is in the .... tense.

Now look at three more examples of the use of the Present Perfect:

*When sufficient liquid has collected in the beaker it should be removed.*
*It will be observed that some of the chalk has not dissolved.*
*Examine the contents of the tube in which the liquids have collected.*

Notice that the Present Perfect occurs in the subordinate clauses (the clauses beginning *when, that, in which*). However, the Present Perfect is only used in such clauses under certain circumstances. The following diagrams give some idea of these circumstances:

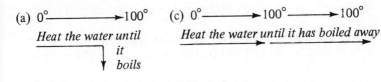

(a) 0°————→100°   (c) 0°————→100°————→100°
*Heat the water until* ... *Heat the water until it has boiled away*

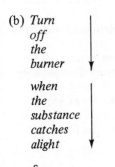

(b) *Turn off the burner*
*when the substance catches alight*

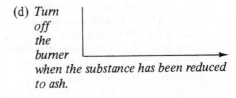

(d) *Turn off the burner when the substance has been reduced to ash.*

G

(a) It boils at a *moment* in time.    Therefore (↓) and use the
(b) It catches alight at a *moment* in time.    Present Simple.
(c) It boils *away* over a *period* of time.    Therefore (→) and use the
(d) It reduces to ash over a *period* of time.    Present Perfect.

Although it is possible to write *Turn off the burner when the substance is reduced to ash*, the Present Perfect gives a more accurate description. This is because it shows that the substance is not immediately reduced to ash.

However, if the length of the period of time is stated, the Present Perfect must be used:

*Turn off the burner when the substance has been left in the flame for three minutes.*
*After they have boiled for five minutes, compare the levels of the liquids.*

Finally, notice that the Present Simple is used if both actions continue for the same period of time:

*Iron expands*
⎯⎯⎯⎯⎯⎯→

*when it is heated*
⎯⎯⎯⎯⎯⎯→

△ **Exercise 10**  Cross out the verb-form in the subordinate clause which you think less good.

1  Most metals contract when they *are cooled/have been cooled.*
2  After it *is used/has been used* several times, it should be thrown away
3  Continue heating until steam *appears/has appeared.*
4  Next, find out the effect of heating on the solution which you *make/ have made.*
5  It will be noted that nothing remains on the filter-paper when the liquid *passes/has passed* through it.
6  Distilling is continued until about 3 cm of the distillate *collects/has collected* in the boiling-tube.
7  Water at 0° emits heat when it *freezes/has frozen* to ice.
8  The gas in the jar should be checked to see whether the soda-lime *does/has done* its work efficiently.
9  When the iron tube *cools/has cooled*, its contents are carefully removed.
10  Pass the magnet over the compound until all the iron filings *are removed/have been removed* from the sand.

11  Leave the litmus paper in the solution until there *is/has been* no further change in colour.

12  At the lowest point of a pendulum swing the potential energy is zero because all the potential energy *is changed/has been changed.*

13  An electric current causes a physical change *when it passes/has passed* through the filament of a light-bulb.

14  When petrol vapour *burns/has burnt* in air there is a rise in temperature.

15  The temperature of the heated water rises steadily until it *reaches/has reached* 100° C. It then remains constant. After it *boils/has boiled* for some minutes, however, it will be noted that some of the water *disappears/has disappeared.*

The Present Perfect is a difficult tense for most learners of English to use correctly. When describing experiments you are advised to stop and think carefully before preferring the Present Perfect to the Present Simple.

**Exercise 11** Write two paragraphs, one describing what happens during *charging*, the other what happens during *discharging*. Use the passive. (The notes may be of help.)

The Lead-Acid Accumulator (charging)

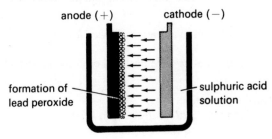

*Charging*

pass current of 2 amps. from (−) plate to (+)
after 10 min. remove charging circuit
remove & examine plates
you find cathode plate → grey
anode plate → brown
brown deposit = lead peroxide
electrical energy put into cell stored in form of peroxide
cell now 'charged'

## *Discharging*

replace plates in acid and connect to torch bulb
bulb lights because chemical energy stored in cell → electrical energy
cell 'accumulator' because accumulates energy
leave lamp connected to cell until—light out
you find almost all brown colour on anode—disappear

## Other uses of the Present Perfect

The Present Perfect is used in certain historical and 'news' statements, but again it is not always easy to use this tense correctly. In fact, as its name suggests, the Present Perfect is only 'half a past tense.' It has been called the 'Pre-Present', and perhaps this is a better name because it is used to describe:

(a) activities carried out a little while before now or just before now.

*A description of the new alloy has recently been published.*
*The research team has just published a description of the new alloy.*
*The properties of semi-conductors have only recently been fully understood.*

Therefore, the present Perfect (or Pre-Present) is usually used with *just* and *recently*. But remember that the Past Simple is used if the statement describes *when* the activity took place:

*A description of the new alloy was published last month.*

(b) activities carried out at some unstated or unspecified time before now:

*Man has been to the moon.*
*A network of communication satellites has been established.*
*The minister has approved the new plan.*

As the above examples suggest, the Present Perfect is used to describe facts. It is used when it seems more important to state the fact that something has been done, rather than when it was done.

(c) activities starting some time in the past, continuing until now, and possibly continuing for some time in the future:

*Medicine has made great progress in the last twenty years.*
*Man has not yet discovered an effective cure for the common cold.*
*This has been known for forty years.*
*This has been known since 1930.*

Compare the last two example sentences with:

*This was first known forty years ago.* (not *for 40 years ago*)
*This was first known in 1930.*

**Exercise 12** Change these sentences into the Present Perfect, using the word or words in brackets. Here is an example:

*This was known forty years ago. (for)*
*This has been known for forty years.*

1  This was known in 1930. (since)
2  The experiment was completed a little while ago. (just)
3  The generator broke down on Monday and Tuesday. (twice this week)
4  The examination finished five minutes ago. (just)
5  The geological survey was published a few weeks ago. (recently)
6  The railway line was closed in 1951. (since)
7  The apparatus was serviced in 1966, 1968 and 1970. (three times so far)
8  The tunnel was closed 80 years ago. (for)
9  This protein was not synthesized last year or this year. (still not)
10  The minister announced last night that the school will be extended (just)
11  Samples of moon dust were brought back by American Astronauts some months ago. (recently)
12  The train arrived a minute ago. (just)

Compare these two pairs of sentences:

*The new library was opened last week.*
*The new library has just been opened.*

*The library was opened ten years ago.*
*The library has been open for ten years.*

The last example sentence shows that the Present Perfect of the main verb *be* can be used with a number of adjectives (like *open*) and prepositional phrases. Study these sentences:

| | | |
|---|---|---|
| *This has* | | *open .....* |
| | | *independent .....* |
| | | *in existence .....* |
| | *been* | *in use/out of use .....* |
| | | *in operation/out of operation .....* |
| *These have* | | *under consideration/under discussion .....* |

Here are some examples:

*Penicillin was first used in 1943.*    *Penicillin has been in use since 1943.*
*The U.N. was founded in 1943.*    *The U.N. has been in existence since 1943.*

*Algeria became independent in 1962.*    *Algeria has been independent since 1962.*
*This has not been used for a year.*    *This has been out of use for a year.*

○ **Exercise 13** Change these sentences in the way shown in the above examples. Remember that *for* is used with a period of time, and *since* is used with a point of time in the past. The first one has already been done.

  1 Penicillin was first used in 1943.
  2 India became independent in 1947.
  3 The Arab league was founded in 1945.
  4 The pipe-line was opened in 1965.
  5 The power-station came into operation in 1966.
  6 The oil-terminal started operations three years ago.
  7 Petrol-driven vehicles were first used about 70 years ago.
  8 Al-azhar University was established in 988.
  9 The Lebanon obtained independence in 1946.
10 The Panama canal was opened in 1914.
11 The old chemical factory was closed four years ago.
12 The problem has been discussed all week.
13 The bridge has not been used since the floods.
14 The salaries of secondary school teachers are being reviewed.
15 This has not worked since the beginning of the year.

△ **Exercise 14** Write out and complete this passage using each of the verbs listed once only.

## The development of world communications

During the last ten years world communications .... very rapidly. An obvious instance .... the sending of communication satellites into orbit. The large increase in the number of messages which .... from one part of the world to another .... the communication explosion. For example, communication satellites .... to send television pictures of the Mexico Olympic Games to countries as far away as Japan and Russia. However, this communication explosion .... to television. The postal services ....; in Western Europe in 1968 nearly a hundred thousand million letters and parcels .... by post.

| | | |
|---|---|---|
| *has been* | *were sent* | *is increasing* |
| *has been called* | *were used* | *are sent* |
| *have developed* | | |
| *have also expanded* | | |
| *has not been restricted* | | |

Read this short passage, paying close attention to the use of tenses.

## The development of domestic heating

*The places where men live have been heated in one way or another ever since the discovery of fire. Originally, the only method of providing heat was by burning wood. Later, charcoal was also used. During the Industrial Revolution, however, coal became available in large quantities, and it was used for domestic heating in many parts of the world. This was because it is a more efficient fuel than wood. But today coal is used less than previously and the main forms of heating are by paraffin stove or by gas or electricity. In cold countries hot water central-heating is also sometimes used.*

Here is a summary of the passage above:

(Domestic Heating)
wood
charcoal
coal
gas, electricity
paraffin
central-heating

**Exercise 15** Write a short passage on one of the following:

1  The Development of Domestic Lighting
2  The Development of Land Transport
3  The Development of Instruments for Measuring Time

You may be able to use some of this information:

| 1 | 2 | 3 |
|---|---|---|
| fire | on foot | sticks |
| oil-lamps | animals | sundials |
| candles | carts | water clocks |
| gas-lamps | trains | mechanical clocks |
| paraffin lamps | cars | watches |
| electric lamps | trucks | electric clocks |
| | | quartz and atomic clocks |

## It + passive verb + that ....

Look at these changes:

*x = y.*
*x = y has been shown.*
*That x = y has been shown.*
*It has been shown that x = y.*

The last sentence is an example of the *it-that* structure. In this structure the *it* is 'empty'; it does not refer to anything. This can be seen from the examples above. The most useful expressions using this structure are as follows:

(a)   *It has been* $\begin{Bmatrix} shown \\ demonstrated \\ proved \end{Bmatrix}$ *that .....*

   *It has therefore been shown that angles ABC and DEF are equal.*

   (This is a usual way of stating impersonally what the writer has done)

(b)   *It will be* $\begin{Bmatrix} seen \\ noted \\ noticed \\ observed \end{Bmatrix}$ *that .....*

   *It will be noted that this result is similar to that obtained by Fisher and Jones.*

   (Although *will* is used, it does not refer to the future. These expressions mean *The reader will see—if he is not stupid—that .....*)

(c)   *It used to be thought that .....*
   *It was once thought that .....*

   *It used to be thought that lead could be changed into gold.*

   (This is a common way of stating that something is no longer believed to be true or possible.)

(d)   *Although it is often said that ...., this is not true.*

   *It is* $\begin{Bmatrix} often \\ commonly \\ generally \end{Bmatrix}$ $\begin{Bmatrix} said \\ believed \end{Bmatrix}$ *that ...., but .....*

*It is commonly believed that acid solutions do not affect gold, but this is not true.*

(This is a way of saying that certain general beliefs are wrong.)

○ **Exercise 16(a)** Complete these sentences putting a suitable *it-that* expression in front. The first one has been done.

1 .... angles *ABC* and *DEF* are equal.
2 .... the earth is flat.
3 .... acids and bases react together to form salts.
4 .... the earth is exactly spherical.
5 .... the temperature is lower at night.
6 .... money always brings happiness.
7 .... the moon was made of cheese.
8 .... magnesium increases in weight when it is burnt.
9 .... atoms are the smallest particles of matter.
10 .... the car travels at a speed of 43 kph.

△ **Exercise 16(b)** Complete the following with sentences of your own. Use different sentences from those used so far.

1 It has now been shown that .....
2 It will be seen that .....
3 It has therefore been proved that .....
4 It used to be said that .....
5 It is often said that .....
6 Although it is commonly believed that .....

# Unit 8  Experimental and explanatory descriptions

Descriptive work plays a large part in most kinds of scientific writing. This unit analyzes and gives practice in some of the most important types of scientific description. These are:

descriptions of experiments
descriptions of how things work
descriptions of how things are produced
descriptions of how things were discovered or invented

## Experimental descriptions

It is not the job of this book to say anything about how experiments should be organised and conducted. However, it can have something to say about how the language used to describe experiments can be organised. In previous units many of the structures commonly used in this kind of writing have already been practised. Now it is time to look more closely at the organisation of individual sentences into paragraphs.

Study this passage:

*Turn a gas-jar upside down and a wooden splint is burnt under it for about a quarter of a minute. I close the jar with a cover and then we put it the right way up on the bench. Next you remove the cover, 2 cm of lime-water is quickly added. Replace the cover and we shake the jar.*

This is part of a description of an experiment to show that $CO_2$ is formed when wood burns. Although the individual sentences are grammatically correct, it is not a good description. (Why?)

The subject-form should not be changed unnecessarily.

Decide which subject-form you are going to use before beginning a description. Here are the five possibilities:

| | | |
|---|---|---|
| (a) Imperative | *Turn a gas-jar upside down.* | |
| (b) Passive | *A gas-jar is turned upside down.* | |
| (c) First person singular | *I turn a gas-jar upside down.* | |
| (d) First person plural | *We turn a gas-jar upside down.* | |
| (e) Second person plural | *You turn a gas-jar upside down.* | |

Although all five forms are possible (a) and (b) are better choices than the others. (c) is usually thought to be too personal. (d) should be used only if the experiment has actually been done by two or more people. (e) is best used for writing instructions, but not descriptions.

Here is the passage rewritten in subject form (a):

*Turn a gas-jar upside down and burn a wooden splint under it for about a quarter of a minute. Close the jar with a cover and then put it the right way up on the bench. Next remove the cover and quickly add 2 cm of lime-water. Replace the cover and shake the jar.*

**Exercise 1** Rewrite the passage using subject-form (b).

**Exercise 2** Rewrite at least one of these grammatically correct descriptions, organizing them better.

## The preparation of water-gas

You soak a small quantity of asbestos wool in water. The asbestos wool is then pushed to the bottom of a heat-resistant test-tube. Then I fill half of the test-tube with small pieces of charcoal. The test-tube is closed with a holed cork. We push a piece of tubing through the cork and an apparatus is set up for collecting the gas over the water. Heat the part of the test-tube containing the charcoal and you collect the gas.

Title: ....

We tie a thread to a metal object which is then suspended from a spring-balance. When the object hangs freely I note the reading on the balance. A graduated cylinder is taken and you then fill it with water up to a convenient point. Place the object in the water in such a way that you cover it completely. Ensure that it hangs vertically and does not come into contact with the walls of the cylinder. We then note the new reading on the spring-balance and the new level in the cylinder.

Now read this description carefully:

A clear glass capillary tube is fixed vertically in a beaker. The beaker contained a liquid which wets glass, such as water or petrol. The liquid rose up the tube and reaches a certain height which depended on the nature of the liquid and the diameter of the tube. With a given liquid, the amount of rise is proportional to the narrowness of the tube. The thickness of the walls of the tube had no effect on the phenomenon. The surface of the liquid at the top of the column was concave to the air.

Again this description is badly organized. (Why?)

The narrative tense should never be changed unnecessarily.

The narrative tense is the tense used to describe the series of operations carried out in the experiment. The narrative tense should be either the Past Simple or the Present Simple, never both in the same description. (However, the special uses of the Present Perfect (Unit 7) should not be forgotten.) But remember that even if the Past Simple is used for the narrative, general statements, such as *water wets glass*, must still be in the Present Simple.

○ **Exercise 3(a)**  Rewrite the description of Capillary Elevation, organisin the tenses. (Choose either Present or Past as narrative.)

□ **Exercise 3(b)**  Write a description of Capillary Depression (i.e., what happens to a liquid which does not wet glass, such as mercury.)

△ **Exercise 4**  Turn this description into the passive using the Past Tense as narrative. (Remember that this does not mean that *all* the verbs should be in the past passive.)

Take a beaker of water and heat it over a burner. Record its temperature every minute. The temperature rises steadily until it reaches 100° but after that it remains constant. Now mark the side of the beaker to indicate the water level. Leave the beaker to boil for several minutes and again record the level. Notice that some of the water has disappeared. The water is changing into water vapour (water vapour from boiling water is called steam). We deduce that the heat being supplied to the boiling water is being used up in the process of turning water into steam. Heat absorbed in this way is known as latent heat.

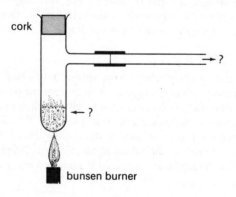

cork

?

?

bunsen burner

△ **Exercise 5(a)** Describe a simple experiment which can be done with the kind of apparatus shown on the last page.

□ **Exercise 5(b)** Describe an experiment which can be done with a more elaborate and complicated version of the apparatus on page 100.

In Unit 4 we saw that most adverbs go immediately before the past participle:

*The phosphorous is carefully dried.*
*Tables are usually made of wood.*

Now notice that in ordinary spoken English we usually find:

*Then the light is turned on.*
*First the water level goes down.*
*Now the answer can be given.*

But in technical written English:

*The circuit is then completed.*
*The water level is first depressed.*
*The answer can now be given.*

A second difference between technical and non-technical writing is that when people write technically they tend to use verbs of Latin origin rather than phrasal verbs* (that is verbs + particles like *take off* or *push down*). Phrasal verbs and Latin verbs belong to different styles. It is therefore better in any one description to keep mainly to one type of verb. Here are a few of many examples of saying (approximately) the same thing:

| *Technical 'Latin' verbs* | *Less technical phrasal verbs* | |
|---|---|---|
| *ignite* | *set fire/light to* | |
| *insert* | *push in* | |
| *depress* | *push down* | |
| *consume* | *use up* | (continued) |

---

* As this book is not primarily concerned with improving comprehension and building up vocabulary, phrasal verbs are only mentioned in passing. Getting students to produce them naturally and correctly is not an easy matter, because phrasal and prepositional verbs fall into a large number of grammatical sub-classes. However, they would certainly be included in any advanced course, or any course aimed at developing spoken technical English.

| *Technical 'Latin' verbs* | *Less technical phrasal verbs* |
|---|---|
| *equalize* | *make up* |
| *occupy* | *fill up* |
| *extinguish* | *put a fire/light out; go out* |
| *add* | *put in* |
| *remove* | *take off/away* |
| *invert* | *turn upside down* |
| *place* | *put* |

(If you look at Exercise 6, you will see how some of these verbs are used; others should be used in Exercise 7.)

The third aspect of the organization of descriptions is rather more difficult because it is a problem of style. Try and avoid putting a 'technical' clause or sentence next to a 'babyish' clause or sentence. Here are three examples of how *not* to write:

*The heat lost by a solid is equal to // how much hotter the water and the container have got.*

*If we do some experiments and some sums we will find that // the velocity V of sound in a given gas is proportional to the square root of the absolute temperature T.*

*If you set a light to a pile of papers, the paper catches fire and bright hot flames eat across the sheets until all the paper has been burnt up. // In such a process a certain amount of energy in the form of both heat and light is emitted.*

The (//) shows that there is a break in style. If possible, try not to have such breaks of style in your own descriptive writing.

**Exercise 6** Look at this experimental description and cross out the alternatives which you think are more babyish, more informal or more likely to be used when speaking rather than writing.

A small $\left\{ \begin{array}{l} \text{bit} \\ \text{piece} \end{array} \right\}$ of phosphorous is carefully dried and $\left\{ \begin{array}{l} \text{placed} \\ \text{put} \end{array} \right\}$ on a

crucible lid inside a bell-jar $\left\{ \begin{array}{l} \text{It is then ignited} \\ \text{Then it is set on fire} \end{array} \right\}$ with a warm glass

rod and a stopper is $\left\{ \begin{array}{l} \text{inserted} \\ \text{pushed in} \end{array} \right\}$ The phosphorous burns producing

$\left\{ \begin{array}{l} \text{clouds} \\ \text{dense white fumes} \end{array} \right\}$ of phosphorous pentoxide which react with the

water. The water level is first depressed as the air $\left\{ \begin{array}{l} \text{becomes warm} \\ \text{gets hot} \end{array} \right\}$

but $\left\{\begin{array}{l}\text{eventually}\\\text{in the end}\end{array}\right\}$ it rises as the oxygen is used up $\left\{\begin{array}{l}\text{As you must make}\\\text{In order to restore}\end{array}\right.$
sure that the pressure left over is normal $\left.\begin{array}{l}\\\text{the pressure of the remaining gas to normal}\end{array}\right\}$ water is now poured into

the trough until the water levels are made equal. Approximately one-fifth of the bell-jar $\left\{\begin{array}{l}\text{is now occupied with water}\\\text{now has water in it}\end{array}\right\}$ showing that one-fifth of the air $\left\{\begin{array}{l}\text{is consumed}\\\text{is used up}\end{array}\right\}$ when phosphorous burns.

**Exercise 7**  Here is an informal description of an experiment. Rewrite it in formal scientific English.

### An experiment to show that carbon dioxide and water are formed when a hydrocarbon burns in air

Set a candle alight and carefully put an upside-down gas-jar over it until the flame goes out. Then take the gas-jar off and put some lime-water in it. Put a cover over the jar and shake it about. The lime-water will turn milky. This shows that there is some carbon dioxide inside. Then let the flame of the candle burn against a cool surface. You will see some drops of a liquid. Put some copper sulphate powder into this liquid and you will see that it turns blue. This shows that the liquid is water.

The last few pages have given practice in writing technical descriptions using the passive. Importance has been given to this kind of writing because it is more difficult. It should not be thought, however, that the passive is always better. Other subject-forms are acceptable as long as they are used consistently.

### Descriptions of how things work

This type of description usually requires a diagram. A diagram makes the explanation easier to follow. A diagram can also be used to avoid the problem of vocabulary. Sometimes you will not know the name for a part of a machine or a piece of apparatus. Never mind! Draw a diagram and label the parts you do not know *A, B, C,* etc.

Read this explanatory description carefully:

*A water tap is a device for turning on and off a flow of water. Its most important parts are a rod with a handle on the top and a washer which is fixed to the bottom of the rod. The metal parts of a water tap are usually made of brass because brass resists corrosion. The washer is made of a flexible material such as rubber or plastic.*

## A Water Tap

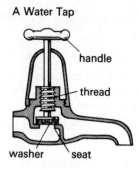

*When the handle is turned the rod either rises or descends because of the spiral thread. The column descends until the washer fits firmly in its 'seat'. (This position is shown in the diagram.) The tap is now closed and no water can flow out of the pipe.*

○ **Exercise 8(a)** Cross out the wrong alternatives. (*S* = sentence)

1 This description consists of 1/2/8 paragraphs.
2 The first paragraph describes a tap/explains how it works.
3 The second paragraph describes a tap/explains how it works.
4 Each paragraph contains 1/3/4/6 sentences.
5 The first sentence (*S*1) is/is not a definition.
6 *S*2 describes the main moving parts of a tap/the main fixed parts.
7 *S*3 explains why brass resists corrosion/why brass is used.
8 *S*4 explains/does not explain why rubber is often used for a washer.
9 *S*5 begins with a subordinate clause/a main clause.
10 *S*6 explains/does not explain why the column goes down.
11 *S*7/*S*8 links the description to the diagram.
12 *S*7 must come before *S*8/it doesn't matter which sentence comes first.

○ **Exercise 8(b)** Write a description of how a water tap works, choosing only five of the eight sentences given in the original passage. In other words, decide which are the five most important sentences and write them out.

○ **Exercise 9** Write a continuous description of how a bicycle pump works choosing one of the given alternatives each time. Your description should therefore contain nine sentences arranged as a passage of continuous English.

## A Bicycle Pump

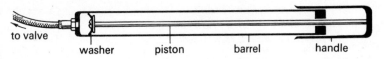

to valve

washer      piston      barrel      handle

1 A bicycle pump is a device $\begin{cases} \text{for forcing water through a narrow tube.} \\ \text{for extracting air from tyres.} \\ \text{for moving air against a pressure difference.} \end{cases}$

2 It $\begin{cases} \text{can} \\ \text{cannot} \\ \text{might} \end{cases}$ work without the valve in the bicycle tyre.

3 $\begin{cases} \text{Essentially,} \\ \text{Firstly,} \\ \text{Importantly,} \end{cases}$ it consists of a hollow barrel, a piston with a handle, and a leather washer at the end of the piston.

4 If the piston is left at the bottom of the barrel the pressure is approximately equal $\begin{cases} \text{that of the atmosphere.} \\ \text{to that of the atmosphere.} \\ \text{to that of the atmospheric.} \end{cases}$

5 When the piston is drawn sharply upwards the air below the piston rises, thus causing the pressure $\begin{cases} \text{to fall} \\ \text{to rise} \\ \text{to remain constant.} \end{cases}$

6 Atmospheric pressure then pushes the sides of the leather washer $\begin{cases} \text{away from} \\ \text{against} \\ \text{through} \end{cases}$ the barrel, allowing air from outside to enter.

7 When the handle is pushed down the air pressure below the piston $\begin{cases} \text{is rising.} \\ \text{rose.} \\ \text{rises.} \end{cases}$

8 This pressure forces the sides of the soft leather washer against the sides of the barrel, $\begin{cases} \text{stopping air from entering.} \\ \text{stopping air from escaping.} \\ \text{allowing air to escape.} \end{cases}$

9 The air is then pumped $\begin{cases} \text{through the tyre-valve into the tyre.} \\ \text{through the tyre into the tyre-valve.} \\ \text{by the tyre-valve into the tyre.} \end{cases}$

H

Notice that three of the nine sentences end with an *-ing* clause:

5 *.... the piston rises, thus causing .....*
6 *.... pushes .... the barrel, allowing air from outside to enter.*
8 *.... forces .... the barrel, stopping air .....*

Finally, here is an example from the last section:

*The phosphorous burns, producing dense white fumes of phosphorous pentoxide.*

These are *-ing* clauses of result:

*x* happens, causing *y* to happen.

Such *-ing* clauses are particularly useful in descriptions of how things work, because with them we can avoid describing a series of events using a series of 'ands':

*x* happens, and *y* happens, and then *z* happens.

A typical sentence structure is subordinate clause + main clause + *-ing* clause. (5) is an example of this:

(a) (subordinate clause)    *When the piston is drawn sharply upwards,*
(b) (main clause)    *the air below the piston rises,*
(c) (*-ing* clause)    *thus causing the pressure to fall.*

○ **Exercise 10** Here are the mixed-up parts of ten (a) + (b) + (c) sentences of this type. Join them together correctly. The first one has been done.

(a)

When the piston is drawn sharply upwards,
When the oven rises in temperature,
As the oven cools,
When the mixture is ignited,
If a bubble of air is introduced into a barometer,
If one end of a metal bridge is fixed to the ground,
When water is heated from 0° C,
If there is a good head of water,
As a rivet cools,
When plates of copper and zinc are placed in dilute sulphuric acid,

(b)

the zinc reacts with the acid,
it pushes the mercury down,

the invar rod is pulled back,
the gas is re-admitted,
the air below the piston rises,
it contracts,
the combustion forces down the cylinder,
the turbine will rotate at high speed,
the other usually rests on a roller,
it contracts,

(c)

thus producing energy.
making the instrument inaccurate.
reaching its maximum density at 4° C.
so cutting off the gas.
drawing the two plates together.
causing the crankshaft to turn.
thus generating large quantities of electricity.
raising the temperature.
thus allowing the bridge to alter its length.
thus causing the pressure to fall.

   As some of these (c) clauses show, the 'result' nature of the *-ing*
clause can be emphasized by putting *so, thus* or *thereby* at the beginning.

△ **Exercise 11** Complete as many of these sentences as you can, by
writing main clauses of your own.

1 ...., causing the water to condense.
2 ...., thus causing the bell to ring.
3 ...., producing a spark.
4 ...., thereby showing that the solution is acidic.
5 ...., causing the vehicle to lose speed.
6 ...., thus controlling the speed of the engine.
7 ...., indicating that a chemical change has taken place.
8 ...., so breaking the current.
9 ...., thereby forcing the rocket into the air.
10 ...., showing that the water molecules pass across the membrane
   into the sugar solution.

△ **Exercise 12** Write a simple explanatory description of one of the
following (diagrams may be used):

| | | |
|---|---|---|
| 1  a bunsen burner | 3  a bus | 5  a fountain pen |
| 2  a burette | 4  an electric switch | 6  a thermometer |

☐ **Exercise 13** Write a more detailed description of how one of the following works (diagrams should be used):

1 an electric bell
2 one type of galvanometer
3 a telescope
4 a thermostat
5 a thermos flask
6 a wind-mill

## How things are produced

We now come to explanations of industrial processes; how substances are purified, how minerals are extracted, how metals and alloys are produced, how objects and materials are manufactured.
    Study this description:

### Sulphur extraction by the Frasch process

*In some parts of the world sulphur deposits lie too deep to be mined in the ordinary way. However, in about 1900 an American engineer called Herman Frasch developed a process for the extraction of this deep-lying sulphur. The Frasch process depends on the fact that the melting point of sulphur is only a little above the boiling point of water. The process consists of three basic operations. First, large amounts of water are super-heated; in other words, the water is heated under pressure to above its normal boiling point. Secondly, this super-heated water is pumped down the well so that it melts the sulphur. Finally, the molten sulphur is pumped to the surface.*

△ **Exercise 14** The description contains seven sentences. (S = sentence) Complete the following analysis:

S1 A statement of the problem of sulphur extraction .....
S2 A statement of when the problem .....
S3 A description of the basic principle .....
S4 A statement that there are only three .....
S5 A description of the first .....
S6 A description of the .....
S7 A description .....

We have seen that the description of sulphur extraction is organized into two main parts: a brief scientific and historical introduction, and a summary of the main operations. The description is also organized in another way:

In some parts of the world (*sulphur deposits lie too deep*) to be mined in the ordinary way. However, in about 1900 an American engineer called (*Herman Frasch developed a process*) for the extraction of (*this deep-lying sulphur*). (*The Frasch process*) depends on the fact that the boiling point of sulphur is only a little above the boiling point of water. The process consists of three basic operations. First, (*large amounts of water are super-heated*); in other words, the water is heated under pressure to above its normal boiling point. Secondly, (*this super-heated water*) is pumped down the well so that it (*melts the sulphur*). Finally, (*this molten sulphur*) is pumped to the surface.

As the arrows show, some of the sentences are linked together by what might be called 'key-phrases'. Important information given in one sentence is referred to again in a later sentence. However, in the later sentence the information is put in a less central position. This is shown below:

| Central position | Non-central position |
| --- | --- |
| (The verb phrase carries the information) | (The noun phrase carries the information) |
| S1 *Sulphur deposits lie too deep* | S2 *This deep-lying sulphur* |
| S3 *Herman Frasch developed a process* | S4 *The Frasch process* |
| S5 *Large amounts of water are super-heated* | S6 *This super-heated water* |
| S6 *So that it melts the sulphur* | S7 *This molten sulphur* |

(Notice also the use of *this* to refer back to something already mentioned.)

The use of such 'key-phrases' can do quite a lot to improve a foreigner's written English. This is because 'key-phrases' make descriptions easier to understand, while, at the same time, they avoid repeating the same information in exactly the same way. And these two problems of being clear and not repeating yourself are particularly difficult if English is not your first language.

○ **Exercise 15** Complete this passage. Remember 'key-phrasing.' (Each
.... signifies one word.)

Copper is usually extracted from an ore called copper pyrites.
Copper ........ are minerals containing a considerable proportion of
copper sulphides. The ........ process is known as smelting. Basically, this
........ consists of heating the ........ in a furnace through which a stream
of air is blown. The ........ melts the ore while the oxygen in the ........
combines with the unwanted elements and removes them. For example,
the ........ combine with the oxygen to form sulphur dioxide. Because
........ ........ is a gas it can easily be removed from the ........ .

○ **Exercise 16(a)** Rewrite the following passage using the verbs below.
Use each verb once only.

Pig iron .... from iron ores such as iron carbonate. The extraction
process .... as smelting. First, the ore .... with coke and crushed limestone. This mixture .... into a blast furnace. As the mixture falls into the
furnace it .... a blast of hot air which fires the coke and .... the temperature of the mixture to about 1800° C. As the coke .... some of it ....
with the oxygen in the air to form carbon monoxide. This, in turn, ....
with the iron to form oxide-free iron and carbon dioxide. Some of the
other impurities, such as sulphur, .... by combining chemically with the
limestone.

| | |
|---|---|
| *burns* | *is extracted* |
| *combines* | *is then fed* |
| *meets* | *is known* |
| *raises* | *is mixed* |
| *reacts* | *are removed* |

△ **Exercise 16(b)** Fill in the first column of 'key-phrases'

> *the extraction process*
> *the ore*
> *this mixture*
> *the air (to form ....)*
> *This, (in turn ....)*

In the above passage there are two sentences of the structure:

*x* + verb .... *y* + *to* + verb + *z* (where *to* = *in order to*)

One is: *This, in turn, reacts with the iron to form oxide-free iron and
carbon dioxide.*

What is the other?

Now compare these two sentences:

(a) *Impurities are eliminated by passing the liquid through a filter.*
(b) *The liquid is passed through a filter in order to eliminate impurities.*

(a) and (b) are two different ways of saying similar things. But they are not saying the same thing. Remember that the subject is a 'strong' position in a sentence. Therefore, a different subject will produce a slightly different description. (a) for instance is principally about *impurities*, while (b) is principally about *the liquid*. In this way it is not difficult to see that:

(a) explains how the impurities are eliminated.
(b) explains why the liquid is passed through the filter.

Now compare:

(c) *CO is formed by combining the burning coke with oxygen.*
(d) *The burning coke combines with the oxygen to form CO.*

Is (c) or (d) the better way of describing part of the iron-smelting process?

As the main purpose of iron-smelting is not to produce *CO*, (d) is a much better descriptive statement. In (c) too much importance is given to how *CO* is formed.
Sentences like (b) and (d) are often called statements of purpose.

**Exercise 17**  Read this description:

## The Treatment of water

Modern methods of treating water usually involve three stages. First, impurities are allowed to settle to the bottom by storing the water in reservoirs. This settling period also kills about 90% of the bacteria. Then the smaller solids and more of the bacteria are removed by filtering the water through sand and gravel. Finally, most of the remaining bacteria are killed by adding a minute proportion of chlorine to the water.

This description is not really satisfactory because too much importance is given to impurities, bacteria, etc., and not enough to water. (Consider the title of the passage.) More emphasis should have been given to what was done to the water and why.

Rewrite the description, using statements of purpose; in other words, *water* should be in the 'strong' subject position.

□ **Exercise 18** Write organized outline-plans for descriptions of two of the following. Then turn one of the plans into a descriptive passage.

1  Coal-mining
2  The refining of mineral oil
3  The production of textiles
4  Glass-making
5  The manufacture of paper
6  Making steel in a Bessemer converter
7  Generating electricity from a hydro-electric plant
8  The manufacture of fertilizers

## How things were discovered or invented

Read this explanatory description:

### The discovery of the velocity of light

*In 1676 Römer discovered that light travels at a determined speed. His discovery was based on the fact that a few years previously other astronomers had identified the satellite of the planet Jupiter. It had also been discovered that each satellite took a certain time to make one circuit of the planet. Römer observed the precise time that one of these moons was in a given position in relation to the planet. For six months he noted that this position was reached progressively a little later each day. After six months he found that the satellite reached the pre-determined position just over 16 minutes later than six months previously. For the following six months the times became progressively shorter. Römer explained this phenomenon in the following way. In six months the earth travels exactly half its orbit round the sun. Hence the shortest times were recorded when the earth was nearest Jupiter and the longest when the earth was farthest away, the maximum difference being 2 × 93 million miles (twice the distance from the earth to the sun). Therefore, the light had travelled an extra 186 million miles and it had taken approximately 16 minutes to do so. Römer thus calculated that the velocity of light is just under 190,000 miles a second.*

In many ways descriptions of inventions and discoveries follow the same principles of organization as other kinds of description. However, the use of tenses is likely to be more of a problem. So, answer these questions:

(a) What is the main narrative tense used in the passage?
(b) Which four verbs are in the Past Perfect?

(c) Can you give reasons for using this tense?
(d) Which three verbs are in the Present Simple?
(e) Why are they in the Present Simple?

Notice that this kind of structure:

$$\text{Römer} \begin{cases} discovered \\ observed \\ noted \\ found \\ calculated \end{cases} that \ldots$$

is more frequently used in describing inventions and discoveries than in most other types of scientific writing. (Why?)

It should also be noted that some of the 'rules' for indirect speech do not apply to scientific and technical English. Compare:

(a) *500 years ago many people thought that the earth was flat.*
(b) *300 years ago Newton stated that white light is composite.*
(c) *Römer discovered that light travels at a determinate speed.*

The last two example-sentences contain general truths. Such true general statements usually remain in the Present Simple even in indirect speech.

△ **Exercise 19(a)** Draw a diagram illustrating Römer's reasoning.

△ **Exercise 19(b)** Write definitions of the following:

1 light      4 a mile
2 a light    5 an orbit
3 velocity   6 a satellite

▢ **Exercise 20** *Either* explain how the speed of sound can be measured, *or* describe any one invention or discovery you know about.

Further work on descriptions

The last unit dealt with experimental and explanatory descriptions. Essentially, this kind of writing describes change—for the most part, chemical changes or physical movements. However, in this unit we will look not so much at descriptions of changes, but at descriptions of states; that is, descriptions of classes of things, of substances and their properties, of the physical characteristics of objects, of the meaning of technical terms, and so on.

## General descriptions

General descriptions describe the properties of classes of things; for example, the properties of solids, or liquids, or gases. They often begin with plural statements:

*Elements are substances that cannot be divided chemically into simpler substances.*

Notice that this statement is similar to a definition except that definitions are nearly always singular, as in:

*An element is a substance which cannot be divided chemically into simpler substances.*

Here are some suggestions for organizing general descriptions.

(a) Always begin with a general statement (often of a defining nature).
(b) Follow complicated general statements with examples.
(c) Explain the meaning of certain technical expressions. Here is a simple example:

  *Liquids possess fluidity. In other words, they do not take any definite shape of their own.*

(d) Always move from the simple to the complex.
(e) Leave statements of use (if any) until towards the end.
(f) Do not contrast what you are describing with anything else until you have established clearly what you are describing in the first place.
(g) Remember that key-phrasing often makes a description easier to understand.

**Exercise 1** Study this description and then answer the questions.

## Elements

Elements are substances that cannot be divided chemically into simpler substances. Elements are chemically indivisible because they consist entirely of atoms which have the same number of protons in the nucleus. In fact, the atomic number of an element is determined by the number of protons. The simplest element is hydrogen which has only one proton. One of the most complicated in uranium. Other common elements are oxygen, carbon, aluminium, and iron. However, many common substances are not elements but compounds. Water, for example, is a compound of hydrogen and oxygen. At present more than a hundred elements are known.

1 The description contains .... sentences.
2 The only sentence of more than two clauses is the .... sentence.
3 *that cannot be divided chemically into simpler substances* is one of the two relative clauses in the passage. The other is .....
4 *cannot be divided chemically → chemically indivisible* is an example of .....
5 There are .... example-sentences in the passage. (An example-sentence does not necessarily contain the words *for example*.)
6 The narrative tense used is the .... tense.
7 This tense is used because .....
8 The .... sentence introduces a contrast.
9 It contrasts .....
10 After *The simplest element is hydrogen, which has only one proton*, it would be very possible to have a sentence beginning, *Therefore*, .....

**Exercise 2** The sentences of the following descriptions have been put in the wrong order. Write the descriptions as continuous English, putting the sentences in the right order.

## Liquids

S1 However, they always maintain their total volume.
S2 A liter of water will remain a liter of water, whether it is poured into a flat dish or into a tall narrow jar.
S3 In other words, they do not take any definite shape of their own.
S4 Liquids possess fluidity.
S5 They immediately take the shape of the container into which they are poured.

## *Atoms*

*S*1 They consist of a nucleus plus one or more electrons.

*S*2 The simplest atom is the hydrogen atom.

*S*3 Atoms are the smallest particles of matter which can take part in chemical reactions.

*S*4 The nucleus consists of one solitary atomic particle called a proton.

*S*5 This atom consists of a nucleus with only one electron revolving around it.

○ **Exercise 3** Write out a continuous description of alloys, choosing one alternative each time. (The sentences are in the right order.)

*S*1 Alloys are metallic substances composed $\begin{Bmatrix} by \\ of \\ from \end{Bmatrix}$ two or more elements.

*S*2 At last one of the elements must be $\begin{Bmatrix} \text{a solid.} \\ \text{a rock.} \\ \text{a metal.} \end{Bmatrix}$

*S*3 Standard steel is an example of an alloy of a $\begin{Bmatrix} \text{metallic} \\ \text{non-metallic} \end{Bmatrix}$ element (iron), and a $\begin{Bmatrix} \text{metallic} \\ \text{non-metallic} \end{Bmatrix}$ element (carbon).

*S*4 Usually, $\begin{Bmatrix} \text{in other words,} \\ \text{however,} \\ \text{therefore,} \end{Bmatrix}$ alloys consist of two or more metal elements.

*S*5 A common example is $\begin{Bmatrix} \text{bronze} \\ \text{invar} \\ \text{brass} \end{Bmatrix}$ which is an alloy of copper and zinc.

*S*6 A rarer alloy is gunmetal, which contains approximately

$\begin{Bmatrix} 90\% \\ 8\% \end{Bmatrix}$ copper $\begin{Bmatrix} 8\% \\ 80\% \end{Bmatrix}$ tin and $\begin{Bmatrix} 2\% \\ 12\% \end{Bmatrix}$ zinc.

*S*7 Alloys are widely used because they often possess more useful properties than $\begin{Bmatrix} \text{pure} \\ \text{impure} \\ \text{solid} \end{Bmatrix}$ metals.

*S*8 For instance, they frequently have greater $\begin{Bmatrix} \text{strength} \\ \text{strong} \end{Bmatrix}$ and $\begin{Bmatrix} \text{hard} \\ \text{hardness} \end{Bmatrix}$ .

Notice that it is possible to write both:

*A common example of a container is a box.*
*A box is a common example of a container.*

Therefore, if you are going to give many examples it is a good idea to move from one structure to the other.

**Exercise 4** Write a general description of hydrocarbons. The following notes should be of help. ($S$ = sentence)

$S$1 hydrocarbons—chemical compounds of carbon and hydrogen
$S$2 simple hydrocarbons—gases
$S$3 the simplest—methane $(C_1H_4)$
$S$4 more complex H-cs—liquid at normal pressure and temperature
$S$5 examples—Benzene $(C_6)$ and Octane $(C_8)$
$S$6 most complicated H-cs $(C_{25+})$—semi-solid
$S$7 e.g., bitumen

**Exercise 5** Write a general description of one of the following:

| | |
|---|---|
| 1 buildings | 6 isotopes |
| 2 clouds | 7 metals |
| 3 detergents | 8 paraffins |
| 4 engines | 9 thermometers |
| 5 gases | 10 waves |

## Descriptions of objects and substances

The organization of this kind of description is similar to that of general descriptions except that

(a) the opening general statements are usually singular.
(b) there is usually no place for examples.

Read this description.

### The planet Mercury

*Mercury is the smallest of the nine major planets. It is also the nearest planet to the sun, its average distance from the sun being approximately 57.5 million kilometers. It takes only 88 days to complete one revolution. Its mass is one-twentyninth that of the earth. It has no atmosphere. It probably does not rotate, so that the side facing the sun is extremely hot and the hidden side is extremely cold.*

△ **Exercise 6** Write a similar description of the planet Jupiter, using the information given below in note form.

Jupiter—largest planet
5th away from the sun (average distance = 773 million km)
one revolution = 11·86 earth years
mass = earth's mass x 317
atmosphere—largely hydrogen
probable temperature at least 150° C.

Read this description.

## Nitric acid

*Nitric acid is a colourless, fuming liquid with a boiling point of 80° C.
It used to be called 'aqua fortis'. It has the chemical formula $HNO_3$.
Nitric acid is a powerful oxidizing agent. It attacks most metals, producing fumes of nitrogen dioxide. Its low boiling point indicates that it is
highly volatile. The reaction of nitric acid with organic substances
produces important compounds such as T.N.T. and celluloid. It is also
widely used in the fertilizer industry.*

△ **Exercise 7** Write a description of sulphuric acid. The following notes
may be helpful.

sulphuric acid—colourless, oily, B.P. 338°
old name—*oil of vitriol*
formula—$H_2SO_4$
extremely corrosive
reacts strongly with water → heat—this property made use of in drying
not volatile
cheapest acid ∴ used in many industries (incl. manufacture of super-
phosphate fertilizers, rayon, detergents and explosives)

Notice that the description of nitric acid contains one statement that
is in the Past Tense—*It used to be called 'aqua fortis.'* This is a special
tense referring to events which occurred for a period in the past but no
longer occur. *Used to* (like *have to* and *ought to*) is an auxilary verb
form. It is quite different to the main verb *to use*. Here are some more
examples:

*Molecules used to be called compound atoms.*
*Mercury used to be called quick-silver.*
*Water pipes used to be made of lead.*
*The removal of the appendix used to be a very serious operation.*

Finally notice the relationship between active and passive in:

*Scientists used to write scientific works in Latin.*
*Scientific works used to be written in Latin.*

Read this description and study the analysis which follows.

## Mercury

*Mercury is a heavy metal which has the unique property of being liquid at normal temperatures. Previously it was mainly used for making mirrors and in the extraction of gold. Today, however, mercury is principally used in the form of compounds in the chemical industry. The main uses of mercury as a metal are in the electrical industry and in thermometers. Nevertheless, mercury is the only metal the production of which has not increased greatly in the last fifty years.*

| Sentence | Time-signal | Tense | Type of information |
|---|---|---|---|
| S1 | (fact) | Present | Definition/properties of mercury |
| S2 | Previously | Past | Previous use |
| S3 | Today | Present | Most important current use |
| S4 | (today) | Present | Other current uses |
| S5 | last 50 years | Present Perfect | Quantity of mercury produced |

△ **Exercise 8** Write a description of copper. If you like, base the description on the following table.

| Sentence | Time-signal | Tense | Type of information |
|---|---|---|---|
| S1 | (fact) | Present | Definition/properties of copper (malleable, 2nd. best electrical conductor, etc.) |
| S2 | The first metal | Past | First metal used by man |
| S3 | For 15,000 years | Present Perfect | In use at least 15,000 years |
| S4 | Today | Present | Most important use (60% as pure in the electrical industry) |
| S5 | (today) | Present | Other main use (40% in alloys) |
| S6 | Recent years | Present Perfect | Pattern of above uses unchanged |

☐ **Exercise 9** Write a description of one of the following:

1  the moon
2  hydrochloric acid
3  aluminium
4  any object or substance you know about

## Descriptions of concepts

These descriptions are usually quite short, being composed of only three or four sentences. However, the structure of the sentences themselves is sometimes rather complicated. This can be seen in the following example.

### *Efficiency*

*Efficiency is the ratio between the energy put into a machine and the energy got out of it. For instance, if falling water delivers 100 horse-power to a turbine and the turbine produces 86 horse-power, then the efficiency of the turbine is 86%. The efficiency of an engine or machine can never be greater than 100%.*

Essentially, descriptions of concepts consist of definitions plus examples. An example is usually included because the concept may be difficult to understand without one. Notice, however, that the example is usually in the form of a conditional sentence. Notice also that the present tense is used in the main clause as well as in the *if*-clause:

*If*
*Suppose that* } .... Present Tense, *(then)* .... Present Tense.

*If falling water delivers ...., then the efficiency is 86%*

○ **Exercise 10** Write out these descriptions and complete them by giving examples. In 1 use a different example to the one already given.

1  Efficiency is the ratio between the energy put into a machine and the energy got out of it. If, for instance, .....
2  Pressure is the force acting on the surface of a body. It is measured as weight per unit area. Suppose, for example, that .....
3  Speed is the ratio of the distance travelled by a moving body to the time taken. If, for example, .....
4  Volume is the measure of the amount of space occupied by a body. The volume of regular solids can be found by measurement of their three linear dimensions. If, for instance, ....    . Liquids, however, ....

Other methods are used to calculate the volume of irregular solids. Suppose that .....

It has been said that descriptions of concepts begin with definitions. Although this is true, the definitions are often in a slightly different form. Compare the following:

| | |
|---|---|
| definition formula | *Efficiency is the ratio of the output energy of a machine to the input energy.* |
| descriptive form | *The efficiency of a machine is the ratio of the output energy of that machine to the input energy.* |
| definition formula | *Speed is the ratio of the distance travelled by a moving body to the time taken.* |
| descriptive form | *The speed of a moving body is the ratio of the distance travelled by that moving body to the time taken.* |

(Notice how *that* is used for the second reference.)

**Exercise 11** Rewrite the following sentences putting them into the descriptive form. The first one has already been done.

1 Speed is the ratio of the distance travelled by a moving body to the time taken.
2 Volume is the measure of the amount of space occupied by a body.
3 A tangent is a straight line touching a curve at one point.
4 Weight is the force of the attraction of the earth on a given mass.
5 Mass is the quantity of matter in a body.
6 Area is the measure of the surface of a body.
7 Refraction is the bending of a ray of light when it passes from one medium to another.
8 $R = \frac{l}{A} \times S$ (1 = length, $A$ = area of cross-section, $S$ = specific resistance of the material)

**Exercise 12** Write descriptions of two of the following.

| | | | |
|---|---|---|---|
| 1 | density | 6 | viscosity |
| 2 | velocity | 7 | stress |
| 3 | voltage | 8 | centrifugal force |
| 4 | valency | 9 | an equation |
| 5 | friction | 10 | the coefficient of expansion |

ɪ

## Descriptions of figures, shapes, and plans

○ **Exercise 13(a)** Study the description of a circle and then complete the sentences below.

### A circle

A circle is a plane geometric figure. It is a closed curve such that the center of the circle is equidistant from all points of the curve. The distance from the center of the circle to the cirumference is called 'the radius.' All the radii of a given circle are equal in length. A straight line from any point on the curve through the center point to the opposite side of the curve is 'the diameter.' Hence, all the diameters of a given circle are also equal.

1 This description of a circle consists of .... sentences.
2 The narrative tense used is .....
3 In five sentences the main verb is .....
4 Only the .... has more than one clause.
5 The following words are in quotation marks because their meanings are explained in the passage—.... and .....
6 In the first sentence .... articles are used because it is a general statement.
7 There are several good examples of 'fronting'; two of them are: .... and .....

□ **Exercise 13(b)** Write a description of one of the following:

1  a cylinder
2  a parallelogram
3  a hexagon
4  an equilateral triangle

The description of the appearance of objects or substances can take many forms. One form is quite common, especially in technical writing. Consider these three statements:

*A triangle is a plane figure which has three sides.*
*A triangle is a plane figure with three sides.*
*A triangle is a three-sided plane figure.*

The structure of the first two example-sentences was considered in Unit 6. The structure of the third sentence is new. Here are the main types.

(a) *blue-lined paper*        colour adjective         }
(b) *a thick-walled vessel*    physical shape adjective  }  + past participle

(c) *an acute-angled triangle*　　geometric shape adjective ⎫
(d) *a three-sided figure*　　　　number　　　　　　　　⎬　τ past participle.
(e) *a U-shaped magnet*　　　　letter　　　　　　　　　⎭

**Exercise 14(a)**  Make simple drawings of the following:

1  an acute-angled triangle
2  an irregular five-sided figure
3  an S-shaped tube
4  a flat-headed screw
5  a wide-toothed saw
6  a two-pronged fork
7  a three-legged stool
8  a star-shaped crystal
9  a flat-bottomed ship
10  a four-bladed fan

You will have noticed that the adjectives and past participles have been joined by hyphens. The hyphen is important because its absence can change the meaning. As in:

*blue lined paper* (i.e., blue paper .....)
*blue-lined paper* (i.e., paper .....)

**Exercise 14(b)**  Describe the following. The first one has been done.

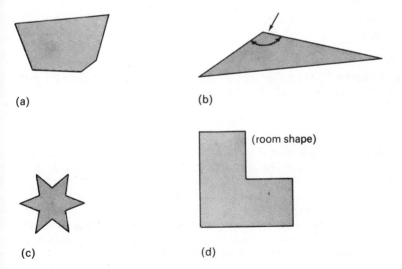

(a)　　　　　　　　　　　　　　(b)

　　　　　　　　　　　　　　　(room shape)

(c)　　　　　　　　　　　　　　(d)

123

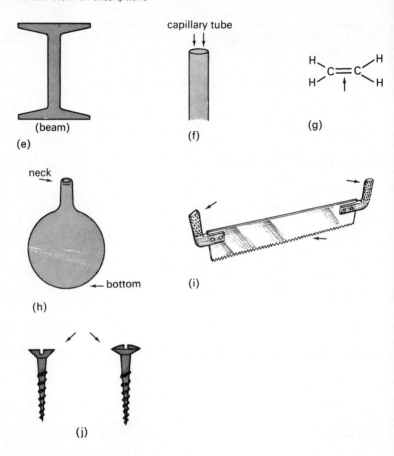

(beam)

(e)

capillary tube

(f)

(g)

neck

bottom

(h)

(i)

(j)

△ **Exercise 15(a)** Study the description and make a drawing of the part of the building described.

## Front elevation

The front of this two-storey building is 9 meters long and 6 meters high. There is a door standing at ground level. The door is 1 meter wide and 2 meters in height and is positioned in the middle. On each side of the door there is a window 1 meter square. The windows are 1 meter off the ground, 1·5 meters from the edge of the door and 1·5 meters from the side of the building. On the second storey there are three identical windows of the same size spaced at intervals of 1·5 meters. The bottom of the windows are 4 meters off the ground.

124

**Exercise 15(b)** Write a description of the part of the building drawn below. Make sure that your description is clear, i.e. that it can be understood in only one way.

Front Elevation

Scale 1 : 100

## Thought-connectives

We have already seen how the distinction between general statements and examples can be made clear by such expressions as *for example* or *for instance*. There are also a number of other words and short phrases that help to make the organization of descriptions clearer to the reader. They have been called 'thought-connectives' because their function is to show the relationship between the thought expressed in one main clause or sentence and the thought expressed in the next main clause or sentance. In previous descriptions these 'thought-connectives' have been used:

*in fact*
*however*
*also*
*therefore*
*hence*
*nevertheless*
*in other words*

As the name 'thought-connective' suggests, these words are rarely necessary in simple physical descriptions. However, they do become useful in more complex scientific and technical writing, and they play an important part in the organization of scientific discussion written in English.

The most important thought-connectives can be grouped in terms of meaning as follows:

(a) *however*
    *nevertheless* } (similar in meaning to *but*)

(b) *thus*
    *therefore*
    *consequently* } (similar in meaning to *therefore*)
    *as a result*
    *hence*

(c) *in fact*
    *in reality* } (used to introduce a contrast with theory; also sometimes used to suggest that the reader is misinformed! e.g., *It is generally thought that x is true. In fact, y is true.*)

(d) *naturally*
    *of course* } (the writer is about to remind the reader of an obvious point
    *obviously*

(e) *in other words* } (the writer is about to state the same thing in another and simpler way)

(f) *as a rule*
    *generally speaking* } (starting a general statement; in contrast to (c))
    *in general*

(g) *then*
    *furthermore*
    *also* } (similar in meaning to *and* but stronger; *and as well, and after that*, etc.)
    *in addition*
    *besides*
    *moreover*

(h) *first*
    *secondly*
    *thirdly* (etc.) } (used when making a series of points. However, if describing the last step in a process, *finally* and not *in conclusion* must be used. Be careful not to use *at first* or *at last*, which are only used to refer to time, i.e., *I waited an hour but at last she came.*)
    *finally*
    *in conclusion*

These meanings are only approximate and the exact use of these words can only be learnt with practice.

○ **Exercise 16** Complete these sentences, putting in suitable thought-connectives.

1 Glass is a very useful material; ...., it breaks easily.

2 Glass is fragile; ...., it should be handled carefully.
3 Glass is made from impure silicon; ...., ordinary sand.
4 Glass is fragile; ...., it has many useful properties.
5 ...., the sand and soda ash are heated to about 1,500° C, but sometimes higher temperatures are necessary.
6 Glass is still made according to principles discovered by the ancient Egyptians; ...., the actual techniques are very different.
7 There are two main reasons for the widespread use of glass bottles; they are cheap and, ...., they have no effect on the flavour, smell or chemistry of the contents.
8 The most remarkable glass production techniques are found in the manufacture of light-bulbs; ...., some machines can produce 7,000 light-bulbs an hour.
9 Liquid glass is viscous; ...., one part will not separate easily from the other parts.
10 Liquid glass is viscous; ...., it is possible to draw molten glass continuously from a tank in order to mass-produce sheets of glass.

There are three important points to notice about thought-connectives and punctuation.

(a) Because thought-connectives join ideas in different sentences or clauses, they must be preceded by stronger punctuation than a comma.

Either a full stop  (..... . *Therefore*, .....)
Or a semi-colon  (.... ; *therefore*, .....)
Or a conjunction  (
and a comma  (.... , *and therefore*, .....)

(b) Thought-connectives are usually followed by a comma.
(c) Thought-connectives usually come first in a clause or sentence.

*However, they can come in other positions.*
*They, however, can come in other positions.*
*They can, however, come in other positions.*
*They can come in other positions, however.*

△ **Exercise 17** Rewrite fifteen of these sentences completing them in any way you like.

1 Liquids possess fluidity; in other words, .....
2 The liquid in this bottle is dangerous; therefore, .....
3 Most kinds of wood float on water; however, .....
4 Aluminium is not as heavy as steel; in fact, .....

127

5 This machine is out of order; as a result, .....
6 This water froze at −1·5° C, but as a rule, .....
7 He said he could run at 60 kph; of course, .....
8 Mercury freezes at −39° C; hence, .....
9 Metals have different properties; in fact, .....
10 Hydrogen is the lightest element; in addition, .....
11 Diamond is the hardest material known; consequently, .....
12 Nitric acid attacked the metal; in general, .....
13 The litmus paper is placed in the acid; as a result, .....
14 An object is multidimensional; in other words, .....
15 The breaking strain was not known; nevertheless, .....
16 The foot-pound system is complicated; obviously, .....
17 Newton was an important scientist; indeed, .....
18 This road is too narrow; furthermore, .....
19 7 is a prime number; .... also .....
20 The properties of an equilateral triangle are, firstly, .....

☐ **Exercise 18** Write an extended description of glass. The description should be organized into paragraphs. Here is a suggestion:

paragraph 1:   The nature and properties of glass
paragraph 2:   The uses of glass
paragraph 3:   The manufacture of glass

(If you want to include historical information it could be put between the first and second paragraphs.)

# Unit 10  More concise statements

## Compound nouns

A compound consists of two or more things joined together. So, a compound noun consists of two or more nouns joined together. For instance, *water* is a noun and *tower* is a noun, so in the sentence, *A new water-tower is being built in the village*, *water-tower* is a compound noun. (On the other hand, there is no compound noun in the sentence, *A tall tower is being built in the village*, because *tall* is an adjective.)

Compound nouns supply information in the shortest or most concise way. Compare the length of these two sentences:

*A new water-tower is being built in the village.*
*A new tower for storing water is being built in the village.*

Compound nouns occur frequently in scientific and technical writing precisely because they give this conciseness. Indeed, the more technical and specialized the subject, the more frequent and the more complicated the compound nouns. Generally speaking, scientific journals contain more compound nouns than university text-books, and university text-books contain more than school books. However, the most complicated of all probably occur in newspaper headlines and technical advertisements, because in both these cases the writers are compelled to put as much information as they can into as few words as possible. Here is an example of each:

*Shoe factory site announcement*
The new low *cost diesel engine transmission unit*

These long compound nouns are sometimes difficult to understand. However, they are not as difficult as they look, if you always follow this principle; begin at the end and then work backwards! Look at this example:

*A day and night weather observation station*

What is the last word? .....
Therefore, the most important thing is that it is some kind of station.
What is the next to last word? .....
Therefore, it is a station for observing things.
What is the third word from the end? ....

Therefore, it is a station for observing the weather.

Now notice that the next part (*day and night*) is linked by *and*, so these words must be taken together. In this way we can see that—

*A day and night weather observation station*

means

*A station for observing weather both by day and by night.*

The idea of beginning at the end may not be easy at first, partly because in many languages complicated nominal groups begin with the most important information and end with the least important.

○ **Exercise 1** Cross out the wrong alternatives.

1  A battery car is } a car driven by batteries.
   a battery for a car.

2  A glass bottle is } a bottle made of glass.
   a kind of glass used in making bottles.

3  A corner house is } a corner of a house.
   a house standing on a corner.

4  A brick floor is } a brick used for building floors.
   a floor made of bricks.

5  A gas turbine is } a turbine which works by gas.
   gas found in or produced by a turbine.

6  A student hostel is } a student who lives in a hostel.
   a hostel for students.

7  An iron pig is } a kind of iron.
   a pig made of iron.

8  A garden flower is } a garden full of flowers.
   a flower found in a garden.

9  A car radio is } a radio for a car.
   a car with a radio.

10  A farm chicken is } a chicken on or from a farm.
    a farm which specializes in producing chickens.

△ **Exercise 2** Write specifying definitions (see page 75) of the following:

1  a car battery        6  a paper filter
2  a battery car        7  satellite communication
3  a race-horse         8  a communication satellite
4  a horse race         9  boat-fishing
5  a filter paper       10  a fishing-boat

There are three things to notice about compound nouns. First, some compounds are joined by a hyphen (-) and some are not. Unfortunately, there are no rules for the use of hyphens. There are differences between one writer and another and, indeed, an individual writer may use a hyphen to join a particular compound on one page and not use a hyphen to join the very same compound on another page. Consider the compound *text book*. This could be written three ways:

*text book*    two separate words
*text-book*    two words joined by a hyphen
*textbook*    one word

Therefore the only thing to do is to copy the forms you find in your reading.

With longer compounds hyphens have to be used with care in order to make the meaning clear. A well-known example is *small car factory*. This could mean either:

a small factory for making cars
a factory for making small cars.

If the second alternative is meant the compound should be written:

*a small-car factory.*

Secondly, notice that *a hostel for students* becomes *a student hostel*. That is, the first half of a compound is never plural. This is because the first noun is in the place of an adjective and adjectives in English are never plural. Also notice that it is still not plural even with a plural number:

*A ruler which measures up to 50 centimeters*
*a 50 centimeter ruler*

○ **Exercise 3** Write out this passage changing the words or phrases in italics into parts of compound nouns.

The box *for the tools, which is made of steel,* is kept in the room *for stores.* In addition to the standard tools, it should contain a ruler *for measuring up to 12 inches,* some wire *made of copper*, some nails *10 centimeters long*, a number of plugs *with two pins*, some wire *for fuses* and a can *containing oil.* If any tools are missing the manager *of the workshop* should be told.

Thirdly, notice that the relationship between the two nouns forming the compound can take many different forms.

131

Study the following:

*a gas-jar*        a jar for containing gas
*a gas-mask*      a mask giving protection against gas
*a gas-meter*     a meter for measuring gas
*a gas company*  a company which supplies gas
*gas fumes*       fumes composed of gas
*gas supply*       the supply of gas
*gaslight*         light supplied by burning gas
*the gas laws*    scientific laws about the behaviour of gases

△ **Exercise 4(a)** Explain the meaning of four of the following:

1  a gas-pipe    4  a gas-oven
2  a gas flame   5  a gas turbine
3  a gas-tap     6  gas-coal

△ **Exercise 4(b)** Explain the meaning of six of the following:

1  an oilcan        6  an oilfield
2  an oil-gauge    7  an oil well
3  an oil tanker    8  an oil expert
4  oil exports     9  oil pollution
5  an oil company  10  an oil heater

The complexity of compound nouns usually makes it impossible to give simple rules for when to use a compound (like *the gas supplies*) and when to use *the*+noun+*of*+noun (like *the supplies of gas*). Choosing the better alternative is something which only comes with practice. To get this practice students should try to use compound forms whenever they think them possible.

However, the difference in meaning between the compound and the noun+*of*+noun structures is clear in one area—that of containers. Consider the difference between:

*a match-box*    *I need a match-box to put this specimen in.*
*a box of matches*  *I need a box of matches to light the fire.*

If there is no container involved there may be no difference in meaning, as in *ice-cube* and *cube of ice*.

△ **Exercise 5** Study the following pairs of expressions. Decide whether they are the same or different in meaning. If you think they are different in meaning, explain the difference.

1  an ink bottle and a bottle of ink

2  a heat source and a source of heat
3  a light-ray and a ray of light
4  a gas-cylinder and a cylinder of gas
5  the world population and the population of the world
6  a cup of tea and a tea-cup
7  a pipe-line and a line of pipes

△ **Exercise 6** Make this passage more concise by using compound nouns.
The parts in italics could be changed.

The size of equipment *used in exchanges for telephones* is being
steadily reduced. The size *of cables* is a recent example. Some of the
latest cables *for telephones* contain 4,800 pairs of wires in a space *of a
diameter of 5·5 centimeters*. This reduction *in weight and cost* has been
achieved by making the material *for insulation, which is made of polystyrene*, extremely thin. However, these cables have reached the specific-
ation *required by the Department of the Post Office*, and they will be
introduced gradually over the period *of the next five years*.

## Naming and the possessive genitive

The possessive genitive is sometimes known as the *'s* genitive. Here are
some examples:

*my father's house*
*John's pencil*
*his uncle's shop*
*the cat's milk*

The possessive genitive is rare in the English used in the physical sciences
and engineering. It is rare because the principal use of the *'s* is to refer
to things possessed by living beings; of course such references are only
rarely made in the physical sciences. It is occasionally used in connec-
tion with non-living things, but in these cases the noun+*of*+noun
structure could always have been used as an alternative. Here are three
typical examples:

*the earth's gravitational field*      *the gravitational field of the earth*
*the moon's surface*      *the surface of the moon*
*the car's maximum speed*      *the maximum speed of the car*

This means that the safest course is never to use the possessive genitive
unless referring to people.

In most scientific writing it is not necessary to state *when* or *where*
somebody discovered or invented something. The discoverer or inventor

is simply named as part of the descriptive statement. Here are two examples:

*A Bunsen burner consists of a metal tube with an adjustable air-valve, used for providing heat in a laboratory.*

*Boyle's law states that at any given temperature the volume of a quantity of gas is inversely proportional to the pressure acting upon the gas.*

Notice that the first example begins with a compound structure:

*a Bunsen burner*

Other examples are: *a Geiger counter, a Leclanché cell, the Leibig method, the Stefan-Boltzmann law*

The second example begins with a possessive genitive:

*Boyle's law*

Other examples are: *Einstein's theory, Clarke's gazelle, Bright's disease*

The question therefore arises; when to use the noun+noun structure and when to use the noun+'s+noun structure. Although there are one or two exceptions these rules should be of help.

Rule 1    References to single people who have made discoveries take the possessive genitive.

Therefore: *Boyle's law, Charles' law, Ohm's law, Fourier's theorem, Avogadro's hypothesis, Bright's disease, Clarke's gazelle*

Rule 2    Reference to single people who have made inventions take the compound noun structure

Therefore: *a Bunsen burner, a Dewar flask, a Fahrenheit thermometer, the Kelvin scale, a Leclanché cell, a Bourdon gauge, a Kohler test, a Rorsarch test*

There remains a middle area in which it is not clear whether something has been invented or discovered. This area includes methods, techniques and reactions. It seems clear, for instance, that both these sentences are equally correct:

*Fulani discovered a new method of staining cells.*
*Fulani invented a new method of staining cells.*

Rule 3    The naming of methods, techniques and reactions may follow either rule 1 or rule 2. It is necessary to learn which form is usually used in each particular case.

Therefore: *Bollman's technique*   but   *the Bechamp technique*
                *Hinsberg's method*   but   *the Liebig method*
                *Kiliani's reaction*   but   *the Sandmayer reaction*

It would seem that the *'s* is used for discoveries that certain things react with other things, while the compound forms are restricted to those reactions which may or may not be very interesting in themselves but have important technical uses in identification and analysis.

Rule 4   References to two or more people (whether inventors or discoverers) require the compound noun structure.

Therefore: *the Stefan-Boltzmann law, the Fischer-Tropsch process, the Joule-Thompson effect*

△ **Exercise 7(a)**   Rewrite, completing as many of the following as possible:

1 Joule's law states that .....
2 Hooke's law states that .....
3 Charles' law states that .....
4 Boyle's law states that .....
5 Ohm's law states that .....
6 Pythagoras' theorem .....
7 Archimedes' principle .....
8 Avogadro's hypothesis .....
9 Carnot's cycle .....
10 Fourier's theorem .....

▢ **Exercise 7(b)**   Describe briefly three of the following:

1 a Farenheit thermometer     6 a Wheatstone bridge
2 a Dewar flask                7 a Leclanche cell
3 a Bunsen burner          8 the Faraday effect
4 the Kelvin scale          9 the Bessemer process
5 a Diesel engine

## Reductions of passive relative clauses

In Unit 5 we saw that sentences containing passive relative clauses could be reduced or shortened. Here are two of the examples given there:

*Pieces of iron which are left in the rain become rusty.*
*An object which is left in the sun becomes hot.*

*Pieces of iron left in the rain become rusty.*
*An object left in the sun becomes hot.*

This type of reduction can be expressed by reduction formula 1:

(art.) + nominal + *wh*-word + *be* + verb-*ed* + prepositional phrase ⟶
 *an       object      which       is        left        in the rain*

(art.) + nominal + verb-*ed* + prepositional phrase
 *an       object      left       in the rain*

Now study these reductions:

| | |
|---|---|
| *Air which is compressed can be used for several purposes.* | *Compressed air can be used for several purposes.* |
| *Add some water which has been distilled.* | *Add some distilled water.* |

This shows that if the passive relative clause consists only of *wh*-word + *be* + verb-*ed* the past participle is placed before the nominal. Hence,

reduction formula 2:

(art.) + nominal + *wh*-word + *be* + verb-*ed*     + φ      ⟶
 φ        air          which        is     compressed   φ

(art.) + verb-*ed* + nominal
 φ       compressed    air

○ **Exercise 8**  Using reduction formula 2, join the following pairs of sentences. The first one has already been done.

1  Air is compressed. It can be used for several purposes.
2  A bottle was broken. It caused the puncture.
3  The metal was heated. It produced certain oxides.
4  Production was improved. It resulted from the reorganization of the factory.
5  He did not obtain the answer. The answer was required.
6  Wires are insulated. They are used to carry the current.
7  Fill three-quarters of a test-tube with water. It should be distilled.
8  All water contains a certain amount of matter. It has been dissolved.
9  Science is a body of knowledge. That knowledge is organized.
10  When chalk is shaken in water it can be seen that some of it does not dissolve. The chalk is precipitated.

It has been said that if a passive relative clause contains only *wh*-word + *be* + verb-*ed* the past participle reduction goes in front of the nominal. This statement is generally true, but it must now be expanded a little. Consider these two reductions:

| | |
|---|---|
| *The liquid which was not wanted was thrown away.* | *The unwanted liquid was thrown away.* |

136

| A bridge which is incorrectly designed may have a short life. | An incorrectly-designed bridge may have a short life. |

(Notice the hyphen (-) in the second reduced form)

This means that reduction formula 2 also applies to passive relative clauses containing *not* or adverbs of manner (i.e., adverbs which describe *how*).

reduction formula 2(a)—negative:

(art.) + nominal + *wh*-word + *be* + *not* + verb-*ed* + φ ————————→
 the     liquid      which      was    not    wanted

(art.) + *un* + verb-*ed* + nominal
 the     unwanted        liquid

reduction formula 2(b)—manner adverb:

(art.) + nominal + *wh*-word + *be* + adverb + verb-*ed* + φ ————————→
 a      bridge      which      is    badly    designed

(art.) + adverb + verb-*ed* + nominal
 a      badly     designed    bridge

o **Exercise 9** Write out the reduced forms of these relative clauses. Use six of the reduced forms in sentences of your own. The first one has already been done.

1 a bridge which is incorrectly designed
2 the substance which is not known
3 cars which are specially built
4 an alloy which is widely used
5 two subjects which are closely related
6 a person who is well informed
7 motion which is uniformly accelerated
8 diagrams which are fully labelled
9 a relationship which is little understood
10 the project which has been half completed
11 conditions which must be completely controlled
12 a problem which has still not been solved
13 workers who are partly skilled } The reductions of these
14 film which has not been exposed enough } three are not straight-
15 film which has been exposed too much } forward.

Now study this reduction:

| Pumps which are driven by wind are sometimes used for raising water. | Wind-driven pumps are sometimes used for raising water. |

κ

In fact, the reduction exemplified by *wind-driven* is not very common and it only seems to be possible with a few verbs such as *drive, make, control, cool, ..... .* It also appears that this reduction can only be used when the agents are of a general nature like *wind, water, man,* etc.

reduction formula 2(c)—general agent:

(art.) + nominal + *wh*-word + *be* + verb-*ed* + $\left\{ \begin{array}{l} by \\ with \end{array} \right\}$ + general noun

$\phi$     pumps     which     are     driven     by         wind

———————➤    (art.) + general noun + verb-*ed* + nominal

              $\phi$      wind      driven      pumps

○ **Exercise 10** Explain in any way you like the meaning of five of the following:

| | |
|---|---|
| 1   a wind-driven pump | 6   man-made fibres |
| 2   a water-cooled engine | 7   chromium-plated steel |
| 3   a hand-painted fabric | 8   frequency-modulated signals |
| 4   factory-made shoes | 9   tin-plated containers |
| 5   a radio-controlled police car | 10   survey-controlled maps |

○ **Exercise 11(a)** Rewrite this passage making it more concise where possible. The italics indicate places where reductions can take place.

The cars *which are used in rallies* like the Safari or the Monte Carlo look the same as ordinary cars *which have been bought from a garage.* In fact they are not the same; they are cars *which have been specially prepared.* Firstly, only components *which have been rigorously tested* are used. Secondly, the components *which are chosen* are put together much more carefully. In other words, the cars *which are driven in rallies* are in effect cars *that have been built by hand.*

○ **Exercise 11(b)** Rewrite this passage making it more concise where possible.

A discovery which was made in 1888 by Hallwachs was the beginning of the study of photo-electricity. Hallwachs took a zinc plate which was highly polished and which was negatively charged. When this plate was illuminated by ultra-violet light it quickly lost its negative charge. On the other hand, a zinc plate which is positively charged does not lose its charge. The loss occurs because light falling on the plate sufficiently energizes the electrons to cause some of them to jump off, carrying their negative charge with them. This phenomenon is the basis of those devices which are widely used which are known as photo-electric cells.

The final expansion of reduction formula 2 involves the two pre-positional phrases *in this way* and *in this manner*. Study this reduction:

| | |
|---|---|
| *The results which were obtained in this way were inaccurate.* | *The results thus obtained were inaccurate.* |
| | *The results so obtained were in accurate.* |

Notice that in this case the past participle comes after the nominal.

reduction formula 2(d)—*so* and *thus*

(art.) + nominal + *wh*-word + *be* + verb-*ed* + $\begin{cases} \textit{in this way} \\ \textit{in this manner} \end{cases}$

  φ  material  which  is  tested  *in this way*

⟶  (art.) + nominal + $\begin{cases} \textit{so} \\ \textit{thus} \end{cases}$ + verb-*ed*

  φ  material  thus  tested

We have seen that passive relative clauses containing only a *wh*-word and a passive verb reduce to the past participle coming just before the nominal. In this way *air which is compressed* reduces to *compressed air*. But we have also seen that the past participle comes after the nominal in the reduction with *so* and *thus*. Now consider:

*A Pitot tube was used to measure the flow through the pipe. The instrument used was type 4CA.*

*The difference between the two weights is the weight of the gas removed from the flask. As the volume of the gas removed can be .....*

*Readings are taken at two-hour intervals. The results obtained are then plotted on a graph.*

These are exceptions to the general rule. Although this part of English grammar is not fully understood, it seems that the past participle only follows the nominal when it refers back to something already explained more fully in previous sentences. This can be seen in the three examples already given. Usually the past participle can come either before of after:

| | | |
|---|---|---|
| *the words underlined* | or | *the underlined words* |
| *the results obtained* | or | *the obtained results* |
| *the sample examined* | or | *the examined sample* |

However, we do *not* find:

*The used instrument was type 4CA.*
*They studied the mistakes. The number of found mistakes was 83.*

○ **Exercise 12** Write out the better alternatives. If you think that both are correct, write out both. The first one has already been done.

1(a) The number of found errors was 103.
 (b) The number of errors found was 103.
2(a) The reason for the crash will probably remain an unsolved problem.
 (b) The reason for the crash will probably remain a problem unsolved.
3(a) Use a covered jar for this experiment.
 (b) Use a jar covered for this experiment.
4(a) Explain the meaning of the underlined words.
 (b) Explain the meaning of the words underlined.
5(a) The early experimenters had only limited resources for making accurate measuring instruments.
 (b) The early experimenters had only resources limited for making accurate measuring instruments.
6(a) In this way radar waves of a known length can be transmitted.
 (b) In this way radar waves of a length known can be transmitted.
7(a) He bought a used car.
 (b) He bought a car used.
8(a) The thieves escaped in a car; the used car was American.
 (b) The thieves escaped in a car; the car used was American.

In Unit 6 this special reduction for the verb *used* was given:

*A pen is an instrument which is* *A pen is an instrument for writing.*
*used for writing.*

Clearly a further reduction is possible:

*A pen is an instrument for writing.* *A pen is a writing instrument.*

reduction formula 3:

(art.) + nominal + *wh*-word + *be* + *used* + *for* + verb-*ing*
 *a*  *tool*  *which* *is* *used* *for* *cutting*   ⟶

(art.) + verb-*ing* + nominal
 *a*  *cutting* *tool*

Notice that these reduced -*ing* verbs are not active. Compare:

*a moving car* (active) i.e., a car which is now moving
*a racing-car* (passive) i.e., a car which is used for racing

The passive reduction usually has a hyphen. It should have a hyphen if there is a chance of misunderstanding.

Therefore, the sentence *Water enters the bottom of the heating flask* could be understood in two ways:

*Water enters the bottom of the flask which is getting hotter.*
*Water enters the bottom of the flask which is used for heating.*

If the second meaning is meant a hyphen should be used:

*Water enters the bottom of the heating-flask.*


**Exercise 13(a)**   Decide whether the words in italics are active or passive in meaning. (The hyphens have been left out!)

1 Most large libraries contain a *smoking* room.
2 The students left the *smoking* room as fast as possible.
3 The police stopped the *racing* car and asked the driver of the old Ford why he was travelling so fast.
4 Many of the latest *racing* cars have 12-cylinder engines.
5 The *rolling* drum escaped down the hill.
6 The *rolling* drum broke down after 500 hours of use.


**Exercise 13(b)**   Describe one from each of the following groups. Diagrams may be used.

1 a sewing-machine, a weighing-machine, a washing machine, a milling machine, a calculating-machine
2 a distilling flask, a deflagrating spoon, a preserving-jar, a tuning fork, a measuring tube, an evaporating basin
3 a reading-room, a waiting-room, a smoking-room, a drawing-board, drawing paper
4 a driving-mirror, a driving licence, a steering wheel, a filling station, a drilling rig
5 a printing press, a rolling mill, a retaining wall, a settling-tank, a softening plant
6 insulating tape, a generating station, recording, a receiving set, deflecting coils


## Reductions of active relative clauses

In Unit 5 this kind of active clause reduction was described:

*This lens produces rays which*      *This lens produces rays converging*
*converge towards a point.*          *towards a point.*

reduction formula 4:

(art.) + nominal + *wh*-word + verb-pres. + prepositional phrase ⟶
 φ        rays        which        converge        towards a point

141

⟶ (art.) + nominal + verb-*ing* + prepositional phrase
    φ    *rays*   *converging*   *towards a point*

If there is no prepositional phrase after the verb, the verb-*ing* usually comes before the nominal.

reduction formula 5:

(art.) + nominal + *wh*-word + verb-pres. + φ ⟶
  φ    *rays*    *which*    *converge*

(art.) + verb-*ing* + nominal
  φ   *converging*  *rays*

Reductions of active relative clauses are rarer than reductions of passive relative clauses. Although most follow formulas 4 and 5 a few do not. For instance:

*The point at which it boils    The boiling point*

There are two more complex reductions of active relative clauses but neither are common. It is probably best for students to use examples they have seen rather than make up their own. Consider this reduction:

*A record which plays for a long    A long-playing record*
*time.*

reduction formula 6:

(art.) + nominal + *wh*-word + verb-pres. + adverb ⟶
  *a*    *river*    *which*    *flows*    *fast*

(art.) + adjective + verb-*ing* + nominal
  *a*    *fast* – *flowing*   *river*

○ **Exercise 14** Apply reduction formula 6 to the following, and then use five of the reductions in sentences of your own. (The last one must be changed somewhat.)

1 a record which plays a long time
2 water which flows fast
3 a drug which acts quickly
4 a submarine which can dive deeply
5 aircraft which fly low
6 bodies which fall freely
7 ground which lies at a low level
8 costs which are always increasing

A more useful reduction (although still restricted) can be seen in:

*Malaysia is a country which*        *Malaysia is a rubber-producing country.*
*produces rubber.*

(As usual, the hyphen can be important. What is the difference between *a glass cutting-tool* and *a glass-cutting tool*?)

reduction formula 7:

(art.) + nominal[1] + *wh*-word + verb-pres. + nominal[2] ──────────►
  a     country    which    produces    rubber

(art.) + nominal[2] + verb-*ing* + nominal[1]
  a    rubber  – producing  country

△ **Exercise 15** Answer as many of these questions as you can. Write complete sentences.

1  Give one important rubber-producing country.
2  Where are some of the world's main wheat-growing areas?
3  What are the most important oil-producing countries?
4  What are the most common food-preserving methods?
5  In what kind of research might a cigarette-smoking machine be used?
6  What is the function of crop-spraying aircraft?
7  What are velocity-retarding forces?
8  Give some examples of earth-moving machinery.

## Reductions of passive 'if' and 'when' clauses

Consider again this pair of sentences:

*An object becomes hot. It is placed in the sun.*

We have already seen that greater conciseness can be obtained by using a reduced relative clause:

*An object placed in the sun becomes hot.*

Now notice that this pair of sentences can be joined equally well by using an *if* or *when* clause:

*An object becomes hot when it is placed in the sun.*
*An object becomes hot if it is placed in the sun.*

Whenever the subject of the main clause and the subject of the subordinate clause refer to the same thing (as in this case) these reductions can be used:

*An object becomes hot when it is placed in the sun.*

*An object becomes hot when placed in the sun.*

*Objects become hot if they are placed in the sun.*

*Objects become hot if placed in the sun.*

reduction formula 8:

$$\left. \begin{array}{l} \textit{When} \\ \textit{If} \end{array} \right\} + \text{nominal} + be + \text{verb-}ed \left. \vphantom{\begin{array}{l} \\ \\ \\ \end{array}} \right\} \quad \longrightarrow$$

$$\textit{When} \qquad it \qquad is \qquad placed \;\; \Big)$$

$$\left. \begin{array}{l} \textit{When} \\ \textit{If} \end{array} \right\} + \text{verb-}ed$$

$$\textit{When} \qquad placed$$

Q **Exercise 16** Using reduction formula 8 complete at least eight of the following. The first one has already been done.

1 An object becomes hot if .....
2 An iron-bar becomes rusty when .....
3 Metals expand when .....
4 Copper contracts if .....
5 The bottle broke into pieces when .....
6 An engine tends to overheat when .....
7 Ebony sinks when .....
8 A bridge may collapse if .....
9 Concrete has greater tensile strength when .....
10 Safety matches only light when .....

Finally notice that the reduced clauses can come at the beginning of the sentence:

*When/If placed in the sunshine, an object becomes hot.*

□ **Exercise 17** Complete eight of the following by supplying suitable main clauses.

1 When placed in the sunshine .....
2 If left on for a long time .....
3 When sent by air .....
4 When tested .....
5 When analyzed .....
6 If heated to above the surrounding temperature .....
7 If required .....

8 If desired .....
9 When previously examined .....
10 When originally planned .....
11 Once opened .....
12 Once ignited .....

# Unit 11 Tables and graphs

Tables and graphs occur frequently in many kinds of scientific and technical writing because they display information in a clear and concise way. However, the information contained in such tables and graphs also usually requires a certain amount of written explanation and discussion. Of course, the written work should not merely describe in detail all the information contained in a table. This would be pointless. In fact, the writer usually wants to pick out the most significant information. He may want to contrast one set of figures with another set, or he may want to compare his results with someone else's.

## Tables and graphs without a time reference

Look at this table:

**Table 1**

**The composition of air** (% volume)

| | |
|---|---|
| Nitrogen | 78·09% |
| Oxygen | 20·95% |
| Argon | 0·9% |
| 5 other gases (neon, etc.) | 0·04% |

Statements from tables with too much detailed information should be avoided. Here is an example of what *not* to do:

*Air is composed of 78·09% nitrogen, 20·95% oxygen, 0·09% argon and 0·04% other gases.*

(In the above example the writer is merely repeating the table.)

Statements with too little information should also be avoided. Here is another example of what *not* to do:

*Air is composed of nitrogen, oxygen, argon and five other gases.*

(The above sentence is merely a list of the constituents of air.)

The written explanation should never merely list the items contained in the table. A long series of items should be divided into a small number of different groups. In the case of Table 1 the written work should be used to point out the differences between the amounts of the various gases. You will find several useful ways of doing this in the following exercise.

○ **Exercise 1** Using the information in Table 1, complete these sentences.

  1 The two most important components of air are .....
  2 The two major components of air .....
  3 The two principal components of air .....
  4 The two main components of air .....
  5 About 4/5 of air is composed of .....
  6 About 1/5 of air is made up of .....
  7 Less than 1 per cent of air consists of .....
  8 Over .... per cent of air is composed of nitrogen and oxygen, of which the former provides easily the .... proportion.
  9 The other five gases occur in only .... proportions.
10 The percentage of air occupied by the noble gases .....

    Now study this table:

**Table 2**

**Composition (by weight) of the earth's crust**

| | |
|---|---|
| Oxygen | 46·5% |
| Silicon | 28·0% |
| Aluminium | 8·0% |
| Iron | 5·0% |
| Calcium | 3·5% |
| Sodium | 3·0% |
| Potassium | 2·5% |
| Magnesium | 2·2% |
| All the other elements | 1·2% |

    It is clear that in terms of per cent occurrence the elements can be placed into a number of different groups. How many groups would you divide the listed elements into? If you chose four groups which elements would you include in each group?

○ **Exercise 2** The following sentences are in the wrong order. Write out an explanation of Table 2 by putting the sentences into the right order.

  1 Next come four elements which are found in percentages varying from 2·2 to 3·5 per cent.

2 Two elements—oxygen and silicon—provide nearly 75 per cent of the weight of the earth's crust.

3 The remaining 95 elements only add up to just over 1 per cent of the total.

4 Just over half of the remaining 25 per cent is provided by the two metallic elements.

5 In fact, only eight elements have a distribution which exceeds 1 per cent of the total.

6 Moreover, the differences between these eight elements are very considerable.

7 The hundred or so elements so far identified exist in widely varying proportions in the earth's crust.

△ **Exercise 3** Describe the melting points of the metals listed in Table 3. (Notice how unhelpful the alphabetical order is.)

**Table 3**

**Melting points (centrigrade) of certain metals**

| Aluminium | 660 |
|---|---|
| Copper | 1083 |
| Iron | 1535 |
| Lead | 327 |
| Platinum | 1773 |
| Tin | 232 |
| Tungsten | 3370 |

You may be able to make use of these kinds of statement:

*The melting points of metals vary* $\left\{ \begin{array}{l} greatly. \\ considerably. \end{array} \right\}$

$x$ *has a* $\left\{ \begin{array}{l} fairly \\ very \\ extremely \end{array} \right\}$ $\left\{ \begin{array}{l} low \\ high \end{array} \right\}$ *melting point.*

$x$ *melts at a* $\left\{ \begin{array}{l} considerably \\ slightly \\ much \\ far \end{array} \right\}$ $\left\{ \begin{array}{l} higher \\ lower \end{array} \right\}$ *temperature than y.*

*With regard to their M.P. these metals can be divided into* .....

□ **Exercise 4** Discuss the information contained in one of the following tables.

### Table 4

**Specific gravity in grams per cc**

| | |
|---|---|
| Aluminium | 2·7 |
| Gold | 19·3 |
| Iron | 7·8 |
| Lead | 11·3 |
| Potassium | 0·8 |
| Sodium | 0·9 |
| Uranium | 19·1 |

### Table 5

**Relative electrical conductivity of pure metals (at 20° C)**

| | |
|---|---|
| Aluminium | 62 |
| Copper | 100 |
| Gold | 72 |
| Iron | 17 |
| Lead | 8 |
| Platinum | 16 |
| Silver | 106 |

### Table 6

**Indices of refraction**

| | |
|---|---|
| Air | 1·0003 |
| Diamond | 2·47 |
| Glass | 1·5−1·72 (approx.) |
| A vacuum | 1·0000 |
| Water | 1·33 |

**Table 7**

**Relative size of the planets in the solar system** (equatorial diameter to nearest 100 miles)

| | |
|---|---|
| Mercury | 3,000 |
| Venus | 7,600 |
| Earth | 7,900 |
| Mars | 4,200 |
| Jupiter | 88,700 |
| Saturn | 74,000 |
| Uranus | 29,000 |
| Neptune | 27,800 |
| Pluto | 3,600 |
| (the sun | 864,000) |

☐ **Exercise 5**  Give a title to the following graph and then write a few lines of explanation.

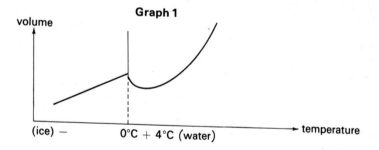

## Tables with a single time reference

Tables with a single time reference describe information given for a certain moment of time. The point of time is usually in the past. This means that the Past tense is usually the narrative tense.

Study this table:

**Table 8**

**Major copper producers (1960)**

| | | |
|---|---|---|
| USA | approx. | 1,050,000 tons |
| Zambia | ,, | 600,000 ,, |
| Chile | ,, | 550,000 ,, |
| USSR | ,, | 500,000 ,, |
| Canada | ,, | 350,000 ,, |
| Congo | ,, | 300,000 ,, |
| Peru | ,, | 120,000 ,, |
| Australia | ,, | 100,000 ,, |
| Japan | ,, | 90,000 ,, |
| Mexico | ,, | 70,000 ,, |

(Total world production ('60) estimated at 4,200,000 tons)

○ **Exercise 6** Complete this explanation of Table 8. Each gap represents one word (or one number).

In 1960 the ten leading copper- .... countries produced nearly .... % of the world total. The .... producer was the USA, one- .... of world production coming from this source. In fact, the USA was the .... country to produce more than .... .... .... . .... other countries each produced half a million tons or more; in decreasing order they were .... Chile and .... . This group was followed by .... countries, Canada and the Congo, producing .... and .... tons respectively. Finally, there .... a group of smaller producers, led by .... and .... . It is worth noting that .... deposits are distributed relatively evenly .... the world because they occur in .... quantities in all continents except .... .

The narrative tense is the Past tense. Why then is the last sentence in the Present Simple?

**Table 9**

**World production of certain metals (1961)**

| | | |
|---|---|---|
| Iron | 190 | million tons |
| Aluminium | 30 | million tons |
| Copper | 4·4 | million tons |
| Zinc | 3 | million tons |
| Lead | 2·5 | million tons |
| Tin | 0·3 | million tons |
| Mercury | 0·09 | million tons |

○ **Exercise 7** Rewrite and complete this passage using information from Table 9. (The spaces represent any number of words.)

As can be seen from .... the world production of iron was .... higher than any .... . In fact, iron production was over .... times as great as its .... competitor, aluminium. Moreover, the latter was easily .... metal with a .... of approximately 30 million tons. Next came a group of .... with a world productions of between 2½ and .... . This group was led by copper, followed by .... . Finally, there were two .... metals listed, the production of which was considerably .... a million tons; they .... .

□ **Exercise 8** Discuss the following table and then compare it with Table 8. (Write at least ten lines.)

**Table 10**

**Major zinc producers (1960)**

| | |
|---|---|
| USA | 450,000 tons |
| Canada | 400,000 tons |
| USSR | 380,000 tons |
| Australia | 270,000 tons |
| Mexico | 270,000 tons |
| Poland | 160,000 tons |
| Peru | 160,000 tons |
| Japan | 150,000 tons |
| Italy | 120,000 tons |
| Congo | 120,000 tons |

## As clauses

Look at these two examples:

*As can be seen from Table 9, the world production of iron in 1961 was not far short of 200 million tons.*

*As is shown in the graph, the relationship between the density and temperature of water is complex.*

This type of *as* clause functions rather like a thought-connective. It connects what is being written to:

| | |
|---|---|
| what has been written already | *As has been stated on page 42, ....* |
| what will be written | *As will be seen, ....* |
| tables, graphs, diagrams, figures or illustrations | *As can be seen on page 8, ....* |

You may have noticed that the structure of these linking *as* clauses is rather unusual in that they have no subjects. Compare these three sentences:

(a) *When it has been proved, the theory will be of practical importance.*
(b) *As has been proved, the theory will be of practical importance.*
(c) *As it has been proved, the theory will be of practical importance.*

What is to be proved in (a)?    *The theory* is to be proved.

What has been proved in (b)?    *That the theory is of practical importance* has been proved.

In other words, the linking *as* clause refers to the whole main clause. This is why it has no subject.

What has been proved in (c)?    *The theory* has been proved.

This shows that (c) is not a linking *as* clause. In (c) *as* means *since*.

Linking *as*-clauses which do not contain modals can be reduced. Here is the formula:

reduction formula 9:

*as* + *be* + verb-*ed* ⟶ *as* + verb-*ed*

| | | | | | |
|---|---|---|---|---|---|
| *As* | *has been* | *stated* | | *As* | *stated* |
| *As* | *has been* | *proved* | | *As* | *proved* |
| *As* | *is* | *shown* | *in Fig. 1* | *As* | *shown*    *in Fig. 1* |

L

△ **Exercise 9**   Complete ten of these sentences. The first one has been done.

  1   As can be seen from Table 9, .....
  2   As can be seen from Table 6, .....
  3   As shown in Table 8, .....
  4   As will be seen from Table 11, .....
  5   As shown in Graph 2, .....
  6   As has been stated at the beginning of this unit, .....
  7   As mentioned in the foot-note on page 23, .....
  8   As can be seen from the figure on page 9, .....
  9   As shown in the following formula, .....
10   As defined on page 69, .....
11   As is well known, .....
12   As has been demonstrated, .....
13   As indicated by the results, .....
14   As expected, .....
15   As can be simply proved, .....

## Tables and graphs with a multiple time reference

The simplest kind of table with a multiple time reference consists of a single column of times or dates against a single column of figures. Look at this example:

**Table 11**

**The growth of scientific journals (approximate figures)**

| Date | No. of Journals in the World |
|------|------------------------------|
| 1750 | 10 |
| 1800 | 100 |
| 1850 | 800 |
| 1900 | 9,000 |
| 1950 | 30,000 |

○ **Exercise 10**   Write out this discussion of Table 9, choosing one alternative each time.

During the last 200 years there $\left\{ \begin{array}{l} \text{was} \\ \text{has been} \\ \text{had been} \end{array} \right\}$ a rapid $\left\{ \begin{array}{l} \text{decrease} \\ \text{increase} \end{array} \right\}$ in the number of scientific journals published. In 1750 $\left\{\begin{array}{l}\text{they}\\\text{there}\\\text{their}\end{array}\right\}$

were only ten in existence, whereas two hundred years $\left\{\begin{array}{l}\text{ago}\\\text{before}\\\text{later}\end{array}\right\}$

30,000 were being published; a rise of $\left\{\begin{array}{l}300\\3{,}000\\300{,}000\end{array}\right\}$ per cent. From 1750

to 1900 the number of journals increases approximately $\left\{\begin{array}{l}2\\10\\30\end{array}\right\}$ times

for each fifty-year period. Although the rate of increase has slowed

down $\left\{\begin{array}{l}\text{since}\\\text{from}\\\text{after}\end{array}\right\}$ then, the number is still increasing $\left\{\begin{array}{l}\text{fastly}\\\text{rapidly}\\\text{rapid}\end{array}\right\}$

Study this graph and the sentences which follow.

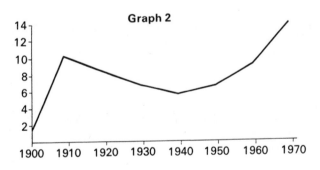

**Graph 2**

1 There was a rise of 9 between 1900 and 1910.
2 There was a fall of 2 between 1910 and 1920.
3 There was a fall of 2 between 1920 and 1930.
4 There was a fall of 1 between 1930 and 1940.
5 There was a rise of 1 during the period 1940 and 1950.

6 There was a rise of 2 over the period 1950–1960.

7 There was a rise of 5 over the period 1960–1970.

8 There was a fall of 11 to 9 and 9 to 7 between 1910 and 1930.

9 There has been a rise of 6 to 7, 7 to 9 and 9 to 14 over the last thirty years.

10 There has been an overall rise of 12 during this century.

△ **Exercise 11** Transform the above sentences, turning them into 'qualified' statements. Here is an example:

5 *There was a very small rise during the period 1940–1950.*

You may be able to use the following:

*See page 160*

$$
\text{There was a (very) }
\begin{Bmatrix}
\text{minimal} \\
\text{small} \\
\text{gradual} \\
\text{slow} \\
\text{large} \\
\text{marked} \\
\text{rapid} \\
\text{sudden}
\end{Bmatrix}
\begin{Bmatrix}
\text{rise} \\
\text{increase} \\
\text{decrease} \\
\text{fall} \\
\text{drop}
\end{Bmatrix}
$$

☐ **Exercise 12** Describe the growth of world copper production over the last 150 years. Use the information given in Table 12. (Be careful about tenses.)

Table 12

**The growth of world copper production (approx. figures)**

| Date | Tons |
|------|------|
| 1810 | 10,000 |
| 1830 | 20,000 |
| 1850 | 50,000 |
| 1870 | 175,000 |
| 1890 | 350,000 |
| 1910 | 930,000 |
| 1930 | 1,175,000 |
| 1950 | 2,600,000 |
| (1960) | (4,200,000) |

156

△ **Exercise 13** Discuss one of the following graphs.

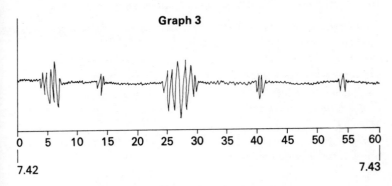

**Graph 3**

A Record of the Sequence and Strength of Earthquake Shocks

For Graph 3 you may find these expressions useful:

the .... shock occurred at ....
the .... shock took place at ....
the .... shock lasted for ....

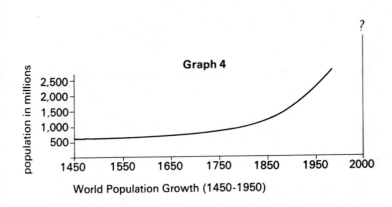

**Graph 4**

World Population Growth (1450-1950)

For Graph 4:

a steady increase .....
a dramatic rise .....
it can be estimated that the world population in 2000 will .....˙

157

So far the tables have contained only a single column of figures. More complicated tables (containing two or more columns) usually involve statements of contrast. Here are three examples:

*The Congo produces more copper than zinc, whereas Canada produces more zinc than copper.*

*Cars A and C have a maximum speed of under 140 kph, while cars B and D have a maximun speed of over 140 kph.* (see page 27)

*Student Group 1 spent more time in classes than in private study; Group 2, on the other hand, spent more time studying privately.* (see page 28)

*Group 1 spent more time on hobbies than in travelling; in contrast, Group 2 spent more time travelling.*

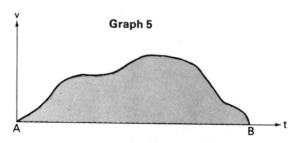

**Graph 5**

Velocity-time Graph of Train Travelling between Station A and Station B

○ **Exercise 14(a)** Complete these statements of contrast, using Table 13.

1 Rain falls every month in London, whereas in Alexandria .....
2 London has six months in a year with an average rainfall of more than 55 mm; on the other hand, Alexandria .....
3 Alexandria usually has three months without rain, while .....
4 Both cities have approximately the same amount of rainfall in January; in March, however, .....
5 London rainfall tends to remain fairly constant throughout the year, whereas .....

**Table 13**

**Average monthly rainfall in millimeters**

|  | Alexandria | London |
|---|---|---|
| January | 48 mm | 47 mm |
| February | 24 | 39 |
| March | 11 | 47 |
| April | 3 | 38 |
| May | 2 | 41 |
| June | 0 | 52 |
| July | 0 | 63 |
| August | 0 | 61 |
| September | 1 | 45 |
| October | 6 | 68 |
| November | 33 | 60 |
| December | 56 | 61 |
| Total | 184 | 622 |

☐ **Exercise 14(b)**   Write a short passage comparing the rainfall of Alexandria and Tripoli, Libya. The information for Tripoli can be found under *City A* in Exercise 14 of Unit 3 (page 33).

☐ **Exercise 15**   Compare and contrast the growth of natural and synthetic rubber. Use Table 14.

**Table 14**

**Growth of world rubber consumption** (in thousands of tons)

|  | 1890 | 1900 | 1910 | 1920 | 1930 | 1940 | 1950 | 1960 |
|---|---|---|---|---|---|---|---|---|
| Natural Rubber | 85 | 170 | 305 | 480 | 850 | 1100 | 1600 | 2050 |
| Synthetic Rubber | 0 | 0 | 0 | 0 | 0 | 20 | 450 | 1900 |

☐ **Exercise 16** Either make up a table or graph of your own and then discuss it or write a discussion of Graph 6.

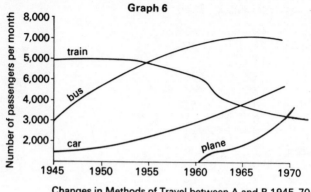

Changes in Methods of Travel between A and B 1945-70

# Unit 12 Reference unit on articles

## The indefinite article

The indefinite articles are *a, an*, and $\phi$. $\phi$ is a special symbol. (For previous uses see the formulas in Unit 10.) This symbol shows that no indefinite article is used. For example, nouns in the plural take $\phi$. In other words, nouns in the plural have no indefinite article. Here is an example:

*Slide-rules are more expensive than ordinary rulers.*

Nouns which can be plural take *a* or *an* in the singular. Here is an example:

*A slide-rule is more expensive than an ordinary ruler.*

*A* is used before a consonant sound, and *an* is used before a vowel sound. Notice that it is the sound which matters, not the letter. This is why we write *an umbrella* but *a university*.

○ **Exercise 1** Put the following into the singular. The first one has been done.

1 Slide-rules are more expensive than ordinary rulers.
2 Apples are round; so are oranges.
3 People who study at universities are called students.
4 Oil wells usually have a greater depth than water wells.
5 Spirit-levels are instruments which work on the principle of an air bubble.
6 Triangles are plane figures which consist of three straight sides.
7 Can atoms be seen with microscopes?
8 Wheatstone bridges are apparatuses for measuring the resistance of electrical circuits.

As Exercise 1 shows, plural nouns never take *a* or *an*. Some non-plural nouns take the indefinite article and some do not. It depends on whether the noun in question is *countable* or *uncountable*. For example, as we can count books we write *a book, two books, three books*, etc. Therefore *book* is a countable noun, as are *an angle, a number, a line*, and *a test-tube*. Countable nouns take *a* or *an* in the singular.

But we can't in English count *bread*. We can't say *a bread, two breads* etc. Therefore *bread* is an uncountable noun, as are *heat, water, power, knowledge*. Uncountable nouns do not take *a* or *an*.

○ **Exercise 2** Cross out the countable nouns in the following lists

1  coffee, oil, year, bottle, wool
2  quantity, wheat, sand, chemistry, point
3  grass, steam, brush, engineering, filter
4  advice, importance, exception, physics, travel
5  information, machinery, equipment, apparatus, training

The following are usually uncountable:

| | | |
|---|---|---|
| names for languages | : | *English, French, Arabic, Hindi, .....* |
| names for subjects | : | *physics, geography, chemistry, .....* |
| names for interests | : | *photography, swimming, stamp-collecting, .....* |
| names for solids | : | *coal, steel, limestone, .....* |
| names for liquids | : | *water, nitric acid, milk, .....* |
| names for gases | : | *air, oxygen, methane, .....* |
| names for powders | : | *salt, sand, sugar, .....* |

These nouns are also usually uncountable:

| | | | |
|---|---|---|---|
| *advice* | *evidence* | *machinery* | *strength* |
| *behaviour* | *furniture* | *produce* | *sunlight* |
| *cardboard* | *grass* | *progress* | *timber* |
| *cloth* | *gravel* | *research* | *traffic* |
| *clothing* | *health* | *rubble* | *training* |
| *concrete* | *heat* | *safety* | *transport* |
| *damage* | *homework* | *scaffolding* | *treatment* |
| *energy* | *information* | *shipping* | *vegetation* |
| *equipment* | *luggage* | *slag* | *work* |

The main danger here is that nouns which are uncountable in English may be equivalent to nouns in the plural in your own language. For example, are your-language equivalents of any of these words ever in the plural in your language?

*advice*
*behaviour*
*damage* (only plural in legal English)
*information*
*furniture*
*research*

It has been said that nouns are either countable or uncountable. Unfortunately, the real situation is not quite so simple as this. There are a number of nouns, especially in scientific English, which can be both countable and uncountable. They will be called 'double' nouns. Sometimes there is a considerable difference in meaning between the countable and uncountable version, sometimes only a small difference. Here are some 'double' nouns, mainly from the scientific and technical fields.

| uncountable | countable |
|---|---|
| *analysis* (in general) | *an analysis* (a particular example) |
| *calculation* (in general) | *a calculation* (a particular example) |
| *diamond* (the hard substance) | *a diamond* (a precious stone) |
| *fertilizer* (in general) | *a fertilizer* (a particular example) |
| *fire* (the general concept) | *a fire* |
| *football* (the game) | *a football* (the ball used in the game) |
| *glass* (the substance) | *a glass* (a container made of glass) |
| *ice* (frozen water) | *an ice* (an ice-cream) |
| *insulation* (in general) | *an insulation* (a particular example) |
| *iron* (*Fe*) | *an iron* (an instrument for ironing) |
| *light* (the general concept) | *a light* (an actual light, often electric) |
| *man* (the general concept, including women and children) | *a man* (an example of the male sex) |
| *paper* (the substance) | *a paper* (a newspaper, a short thesis) |
| *rope* (in general) | *a rope* (a piece of actual rope) |
| *rubber* (the substance) | *a rubber* (for rubbing out mistakes, etc.) |
| *stone* (the substance) | *a stone* |
| *wire* (in general) | *a wire* (a piece of actual wire) |
| *wood* (the substance) | *a wood* (a large group of trees) |
| *science* (in general) | *a science* (a particular science) |
| *sound* (in general) | *a sound* (a particular sound) |

△ **Exercise 3** Complete at least eight of these sentences, showing that you have understood the different uses of 'double' nouns.

1 Diamond .....
2 A diamond .....
3 Glass .....
4 A glass .....
5 Light .....
6 A light .....
7 Iron .....
8 An iron .....
9 Man .....
10 A man .....
11 Science .....
12 .... a science.

163

Now consider:

*A thermometer measures temperature.*
*Temperature is usually expressed in degrees.*
*A temperature of over 50° Centigrade was recorded.*
*The boy had a high temperature.*

These four example-sentences show that a number of general concepts, of which *temperature* is one, can be countable in certain situations. Principally these are:—

(a) after the verb *have* (*has a temperature* and also see page 16)
(b) before an *of* phrase (*a temperature of 250°*)
(c) before an adjective (*a high temperature*)

Other nouns which follow the pattern of *temperature* are:

| | | |
|---|---|---|
| *pressure* | *mass* | *velocity* |
| *force* | *voltage* | *density* |
| *growth* | *gravity* | *strength* |

Finally, in highly technical writing it is possible to find nouns which are always uncountable in ordinary English becoming uncountable. Here are two examples:

*Cheaper mild steels are now being produced.*
*Improved wheats will be introduced on a wide scale this year.*

There are probably two reasons for this phenomenon. One is the need to be concise. Clearly *improved varieties of wheat* takes up more space than *improved wheats*. The second reason is that a non-expert probably does not see much difference between different kinds of wheat and steel. To an expert, however, such differences are obvious; hence, his use of *steels* and *wheats*.

(You are not advised to use ordinary language uncountables in the plural unless you have already seen specific examples in textbooks.)

## The definite article

If there are certain problems with the correct use of the indefinite article, there are probably more with the definite article (*the*). This is because the definite article is used in two quite different (and indeed opposite) ways in English. However, the main use of the definite article is to specify. In other words, it is used when referring to a particular thing or particular things.

(a) *The* may specify when the person(s) or thing(s) have already been mentioned. Study these examples:

> *They poured the liquid into a beaker. The beaker was then placed in a retort stand. The retort stand was then moved .....*
> *It has a small engine which is in the front. The engine has a capacity of 1,100 cc.*

An exception to this rule is specific definitions (see pages 75–77). In this case the second mention of the noun remains indefinite:

*A key-hole saw is a saw with a narrow blade .....*
*A suspension bridge is a bridge which .....*

(b) *The* may specify when it is obvious who or what is being referred to and there is no chance of mistaken identity. Study these examples:

> *The sun rises in the east.* (is it clear *which* sun is meant?)
> *Information about the weather comes from weather observation stations.* (is it clear *which* planet's weather is meant?)
> *The oxygen balance in the atmosphere is maintained by photosynthesis.* (is it clear *which* atmosphere is meant?)
> *The decimal system is widely used in science.* (How many decimal systems are there—is there any chance of confusion?)

If you can meaningfully ask yourself the question *which one is meant?* then you should use *the*; if not, not.

(c) *The* usually specifies when the noun is followed by an *of* phrase (even if the sentence is a general statement). Look at these examples:

> *The area of a circle equals $2\pi r^2$.*
> *Everybody recognizes the importance of practical work.*
> *The coefficient of expansion of brass is 0·000026 per °C.*
> *The apparent loss of weight of a substance which is immersed in a liquid equals the weight of liquid displaced.*

*The* is usually used to make specific, particular statements. It follows, therefore, that *the* is not usually used in general statements. It is *not* used, for example, in cases like these:

*Water boils at 100° C.*
*Mathematics is the basis of science.*
*Mild steel is used for making wire.*
*Hydrometers measure the specific gravity of liquids.*

But what is a general statement and what is a specific statement? Look at these four sentences:

(a) *They will read books.*
(b) *They will read technical books.*
(c) *They will read books on heat, light and sound.*
(d) *They will read the books on heat, light and sound recommended to them.*

(a) is obviously general—so no *the*. (d) is obviously specific because we can ask, *Which books were recommended?*. But (b) and (c) are also general, because we are still talking generally about books, although this is now restricted to talking generally about certain kinds of books. But if we ask the question *Which books on heat, light and sound?*, the only possible answer is *Any books*. Therefore in (b) and (c) the definite article is not used.

Most languages have something like a definite article, and in most languages the definite article is used in general statements. This means that students have to be particularly careful not to use *the* too much when they write English.

△ **Exercise 4** Read this passage, translate it into your language, and then answer the questions.

Physics is largely concerned with measurement—especially the measurement of physical quantities such as length, time, velocity, volume, mass, density, weight and energy. Many of these quantities are interrelated. Volume is a length multiplied by a second length and then multiplied by a third length. Velocity is distance divided by time. Density is mass divided by volume. In fact, most physical quantities are related to length, time and mass.

1 How many times is the definite article used in the English passage?
2 How many times is the indefinite article used?
3 Why is the indefinite article used so few times?
4 When you translated the passage how many times did you use your-language equivalents of the definite article?
5 What differences are there between the use of the definite article in English and the use of the definite article in your language?

○ **Exercise 5** Rewrite this passage putting in *a, an* or *the* where necessary.

When .... sunlight strikes .... object .... colour of .... object depends upon .... wavelengths which .... object reflects. If, for example, .... grains

of .... sugar reflect equally well all .... wavelengths of .... spectrum ....
grains appear white. If a surface reflects only .... wavelength which
produces red and absorbs .... other waves of .... spectrum .... surface
appears red. Black is .... absence of .... colour because .... black objects
absorb all .... light of .... spectrum.

Up till now, it has been said that *the* is used when we want to refer
to a particular thing or things—this is why it is called the definite article.
Sometimes, however, with a singular countable noun *the* is used in a
general sense. This unspecific or 'universal' *the* can be used when talking
about the general idea of a thing. There are only a few examples of
'universal' *the* in this book, and nearly all of them are in Unit 7. Here
are a few examples:

*The steam-engine was invented in the nineteenth century.*
*The hydrogen bomb is the most dangerous thing that man has invented.*
*The lathe is one of the most important tools in a mechanical work-shop.*
*The camel is being superceded as a means of transporting goods.*

THOMAS NELSON AND SONS LTD
36 Park Street London W1Y 4DE
PO Box 18123 Nairobi Kenya

THOMAS NELSON (AUSTRALIA) LTD
597 Little Collins Street Melbourne 3000

THOMAS NELSON AND SONS (CANADA) LTD
81 Curlew Drive Don Mills Ontario

THOMAS NELSON (NIGERIA) LTD
PO Box 336 Apapa Lagos

THOMAS NELSON AND SONS (SOUTH AFRICA)
(PROPRIETARY) LTD
51 Commissioner Street Johannesburg

0 17 555017 4

Printed in England by
Butler and Tanner Ltd., Frome.

168